GAME THEORY FOR ECONOMIC ANALYSIS

This is a volume in
ECONOMIC THEORY, ECONOMETRICS, AND MATHEMATICAL ECONOMICS

A Series of Monographs and Textbooks

Consulting Editor: KARL SHELL

A complete list of titles in this series appears at the end of this volume.

GAME THEORY FOR ECONOMIC ANALYSIS

Tatsuro Ichiishi

DEPARTMENT OF ECONOMICS
DEPARTMENT OF MATHEMATICS
THE UNIVERSITY OF IOWA
IOWA CITY, IOWA

1983

ACADEMIC PRESS

A Subsidiary of Harcourt Brace Jovanovich, Publishers
New York   London
Paris   San Diego   San Francisco   São Paulo   Sydney   Tokyo   Toronto

COPYRIGHT © 1983, BY ACADEMIC PRESS, INC.
ALL RIGHTS RESERVED.
NO PART OF THIS PUBLICATION MAY BE REPRODUCED OR
TRANSMITTED IN ANY FORM OR BY ANY MEANS, ELECTRONIC
OR MECHANICAL, INCLUDING PHOTOCOPY, RECORDING, OR ANY
INFORMATION STORAGE AND RETRIEVAL SYSTEM, WITHOUT
PERMISSION IN WRITING FROM THE PUBLISHER.

ACADEMIC PRESS, INC.
111 Fifth Avenue, New York, New York 10003

United Kingdom Edition published by
ACADEMIC PRESS, INC. (LONDON) LTD.
24/28 Oval Road, London NW1 7DX

Library of Congress Cataloging in Publication Data

Ichiishi, Tatsuro.
  Game theory for economic analysis.

  (Economic theory, econometrics, and mathematical
economics)
  Bibliography: p.
  Includes indexes.
  1. Game theory. 2. Equilibrium (Economics) I. Title.
II. Series.
HB144.I28 1982   330'.01'5193   82-11554
ISBN 0-12-370180-5

PRINTED IN THE UNITED STATES OF AMERICA

83 84 85 86   9 8 7 6 5 4 3 2 1

To Barbara

Contents

Preface ... ix

0. Preliminary Discussion
0.1. Basic Notation ... 1
0.2. Pure Exchange Economy ... 3

1. Introduction to Convex Analysis
1.1. Convex Set ... 8
1.2. Affine Subspace ... 10
1.3. Hyperplane ... 12
1.4. Algebraic Interior, Algebraic Relative Interior ... 13
1.5. Separation Principle ... 17
1.6. Extreme Points ... 21
Appendix ... 26
Exercises ... 27

2. Introduction to Continuity of a Correspondence
2.1. Upper and Lower Semicontinuities ... 32
2.2. Closedness ... 34
2.3. Maximum Theorem ... 37
Exercises ... 38

3. Introduction to Fixed-Point Theorems in R^n
3.1. Sperner's Lemma, K–K–M Theorem ... 43
3.2. Fixed-Point Theorems ... 47
3.3. Fixed-Point Theorem and Separation Principle: Coincidence Theorem ... 50
Exercises ... 53

4. Noncooperative Behavior and Equilibrium
4.1. Nash Equilibrium of a Game in Normal Form ... 56
4.2. Optimality ... 58
4.3. Social Equilibrium of an Abstract Economy ... 60
4.4. Competitive Equilibrium of a Pure Exchange Economy ... 61

4.5.	Fundamental Theorems of Welfare Economics	62
4.6.	Game-Theoretical Interpretation of the Competitive Equilibrium	65
4.7.	Notes	67
	Exercises	74

5. Cooperative Behavior and Stability

5.1.	Linear Inequalities	78
5.2.	Core of a Side-Payment Game	80
5.3.	K–K–M–S Theorem	82
5.4.	Core of a Non-Side-Payment Game	83
5.5.	Core Allocation of a Pure Exchange Economy	86
5.6.	A Limit Theorem of Cores	89
5.7.	Social Coalitional Equilibrium of a Society	94
5.8.	Optimality of the Nash Equilibrium: Strong Equilibrium	100
5.9.	Notes	102
	Exercises	113

6. Cooperative Behavior and Fairness

6.1.	Shapley Value of a Side-Payment Game	118
6.2.	Convex Game	120
6.3.	λ-Transfer Value of a Non-Side-Payment Game	122
6.4.	Value Allocation of a Pure Exchange Economy	126
6.5.	A Limit Theorem of Value Allocations of Replica Economies	128
6.6.	Notes	145
	Exercises	148

References	151
Author Index	157
Subject Index	161

Preface

The purpose of this book is to present the development over the past three decades of three strands in n-person game theory, with strong emphasis on applications to general economic equilibrium analysis. The first strand, represented by the *Nash equilibrium* of a game in normal form, studies an equilibrium concept of a society in which everybody behaves noncooperatively and passively. The second strand, represented by the *core* of a game in characteristic function form, studies the stable outcomes of a society in which everybody is aware of what he can do by cooperating with other members. The third strand, represented by the *Shapley value* of a side-payment game in characteristic function form, studies the fair outcomes of a society in which everybody is aware of what he can contribute to any group of members of the society by joining the group.

To capture the game-theoretical idea of each strand, several formulations with diverse degrees of generality have been proposed and studied in the past. A simple formulation conveys the essence of the idea in a straightforward manner, but it may not be general enough for useful economic applications. A complex formulation is general and powerful for economic applications, but beginners may find it hard to capture the essence. For each strand, therefore, I first present the game-theoretical idea in a simple setup and then gradually generalize it to more complex situations. This task is followed by discussion showing how the game-theoretical solution concept of each strand, generalized to the appropriate degree, can be applied to general economic equilibrium analysis. The model of a pure exchange economy is chosen, and three existence theorems for this model are established: the competitive equilibrium existence theorem (an application of a generalized Nash equilibrium existence theorem), the core allocation existence theorem (an application of a theorem for nonemptiness of the core), and the value allocation existence theorem (an application of a generalized Shapley value existence theorem). Having thus bridged n-person game theory and economic theory, I proceed to present several issues in mathematical economics dealing with the three economic concepts (competitive equilibrium, core allocation, and value allocation) within the framework of pure exchange economies; in particular, I present the fundamental theorems of welfare economics and limit theorems of cores and of value allocations. I also present a still more general game-theoretical concept, developed recently by myself: a social coalitional equilibrium. I do not discuss its economic applications here, however, because in order to do so I would have to give a disproportionately long discussion.

The text is organized as follows: Chapters 1-3 are devoted to some mathematical tools and theorems, which are usually not covered in standard mathematics courses but which play crucial roles in this text. In Chapter 4 (5, 6, respectively), a systematic account is given for the first (second, third, respectively) strand of n-person game theory mentioned above. Sections 4.4, 5.5, and 6.4 are the bridges between game theory and mathematical economics (a competitive equilibrium existence theorem, a core allocation existence theorem, and a value allocation existence theorem). Most of the later sections of the three chapters deal with mathematical economics. A social coalitional equilibrium is discussed in Sections 5.7 and 5.8. Sections 4.7, 5.9, and 6.6 are bibliographical notes. There, the evolution of relevant concepts is surveyed and more recent results are stated.

The reader is assumed to be familiar with junior-level real analysis and linear algebra. Every effort has been made to make the exposition self-contained, given this mathematical background.

The present work arose out of several courses that I have given since spring 1978 at Carnegie-Mellon University and at The University of Iowa. It was George J. Fix, head of the Mathematics Department, Carnegie-Mellon University, who first offered me an opportunity to develop a course in game theory and mathematical economics. Juan Jorge Schäffer of Carnegie-Mellon University shaped up my mathematical thinking, not only through our delightful collaboration in cooperative game theory, but also through innumerable discussions on mathematical science in general. While the first draft of the text was being typed in spring 1981, some of my colleagues at The University of Iowa, in particular Michael Balch and John Kennan, suggested some improvements in the text. Richard P. McLean of The University of Pennsylvania read the entire first draft and gave me many pieces of valuable information and thoughtful suggestions; indeed, most of the revisions I have made since then originated from his suggestions. My results that are included in the text were established in my research project, supported by the National Science Foundation Grant SES 8104387 (formerly, SOC 78-06123). I would like to thank the staff of Academic Press for the excellent work that was done to produce this book. To all the individuals and the institutions mentioned in this paragraph, I would like to express my deep gratitude. Needless to say, I am solely responsible for any possible deficiencies in this book.

0

Preliminary Discussion

The basic mathematical notation to be used throughout this text is summarized in Section 0.1. Since the game theory presented in Chapters 4–6 serves as *the* mathematical foundation for economic analysis, *the* typical economic model is presented in Section 0.2: the model of a pure exchange economy. The reader is encouraged to keep this model in mind when going through Chapters 4–6.

0.1. Basic Notation

Given a set A,

$\#A := $ the cardinality of A.

Given a positive integer k,

$\mathbf{R}^k := k$-dimensional Euclidean space;
$\mathbf{R}^k_+ := $ the nonnegative orthant of $\mathbf{R}^k$;
$\mathbf{R} := \mathbf{R}^1$;
$\mathbf{R}_+ := \mathbf{R}^1_+$.

For any $x, y \in \mathbf{R}^k$,

$x_i :=$ the ith coordinate of x, $i = 1, \ldots, k$;
$x \cdot y := \sum_{i=1}^{k} x_i y_i \equiv$ the Euclidean inner product of x and y;
$\|x\| := \sqrt{x \cdot x} \equiv$ the Euclidean norm of x;
$x \geq y$ means $x_i \geq y_i$ for every $i = 1, \ldots, k$;
$x > y$ means $x \geq y$ and $x \neq y$;
$x \gg y$ means $x_i > y_i$ for every $i = 1, \ldots, k$.

For any subsets S, T of $\mathbf{R}^k$,

$\overset{\circ}{S} :=$ the interior of S in $\mathbf{R}^k$;
$\bar{S} :=$ the closure of S in $\mathbf{R}^k$;
$S + T := \{x + y \in \mathbf{R}^k \mid x \in S, y \in T\}$;
$S - T := \{x - y \in \mathbf{R}^k \mid x \in S, y \in T\}$.

The algebraic concepts $S + T$ and $S - T$ should not be confused with the set-theoretic concepts $S \cup T$ and $S \setminus T$. Abbreviations for some phrases:

    iff   if and only if;
  w.l.o.g. without loss of generality;
    □   the end of the proof.

A positive integer n will be interpreted as the number of *players* throughout Chapters 4–6. A set of players is called a *coalition*:

$N := \{1, 2, \ldots, n\}$ is interpreted as the set of players;
$\mathcal{N} := 2^N \setminus \{\emptyset\}$ is interpreted as the family of nonempty coalitions.

For every $j \in N$

$$e^j := (0, \ldots, \overset{j}{1}, \ldots, 0) \in \mathbf{R}^n.$$

For every $S \in \mathcal{N}$,

$\chi_S := \sum_{j \in S} e^j$;
$\Delta^S :=$ the convex hull of $\{e^j \mid j \in S\}$ (see Section 1.1).

Let X be a convex subset of $\mathbf{R}^k$, and let $f : X \to \mathbf{R}$ be a function. The function f is called *quasi-concave in* X if for every $r \in \mathbf{R}$ the set

$\{x \in X \mid f(x) \geq r\}$ is convex or, equivalently, if for any $x, y \in X$ and any t in the unit interval $[0, 1]$ it follows that $f(tx + (1 - t)y) \geq \min[f(x), f(y)]$. The function f is called *concave in X* if for any $x, y \in X$ and any $t \in [0, 1]$, $f(tx + (1 - t)y) \geq tf(x) + (1 - t)f(y)$. It is called *strictly concave in X* if strict inequality holds true in the last inequality whenever x and y are distinct and $0 < t < 1$.

0.2. Pure Exchange Economy

The model of a pure exchange economy with l types of commodities and m consumers is reviewed. A *commodity bundle* is a point x in $\mathbf{R}^l$; it describes the quantity x_h of each commodity $h = 1, \ldots, l$. The ith consumer is characterized by a triple $(X^i, \lesssim_i, \omega^i)$ of his consumption set X^i, his preference relation $\lesssim_i$, and his initial endowment vector ω^i. The *consumption set* X^i is a subset of $\mathbf{R}^l$ and is interpreted as the set of all commodity bundles with which he can physically survive; the set X^i characterizes the "physical needs" of consumer i. Suppose he chooses a bundle $x^i \in X^i$. If $x_h^i > 0$ ($x_h^i < 0$, resp.), then he demands (supplies, resp.) $|x_h^i|$ units of commodity h. The *preference relation* $\lesssim_i$ is a binary relation on X^i. The statement $[x^i \lesssim_i x^{i'}]$ is interpreted as: the commodity bundle $x^{i'}$ is at least as desirable as the commodity bundle x^i to consumer i. The relation $\lesssim_i$, therefore, characterizes the "taste" of consumer i. Given $x^i, x^{i'} \in X^i$, denote by $[x^i >_i x^{i'}]$ the negation of $[x^i \lesssim_i x^{i'}]$. Denote by $[x^i \sim_i x^{i'}]$ the statement $[x^i \lesssim_i x^{i'}, \text{ and } x^{i'} \lesssim_i x^i]$, that is interpreted as: Consumer i is indifferent as to the choice of commodity bundle x^i or commodity bundle $x^{i'}$. The preference relation $\lesssim_i$ is called *complete* if for any $x^i, x^{i'} \in X^i$ it follows that $x^i \lesssim_i x^{i'}$ or $x^{i'} \lesssim_i x^i$; completeness means that consumer i has a strong opinion on the commodity bundles. It is called *transitive* if for any $x^i, x^{i'}, x^{i'''} \in X^i$ for which $x^i \lesssim_i x^{i'}$ and $x^{i'} \lesssim_i x^{i'''}$ it follows that $x^i \lesssim_i x^{i'''}$; transitivity means that the consumer is rational. It is called *closed* if for any $x^i \in X^i$ the sets $\{\xi^i \in X^i \mid \xi^i \lesssim_i x^i\}$ and $\{\xi^i \in X^i \mid x^i \lesssim_i \xi^i\}$ are both closed in X^i; closedness means that his comparison of commodity bundles is smooth. It is called *weakly convex* if for any $x^i \in X^i$ the set $\{\xi^i \in X^i \mid x^i \lesssim_i \xi^i\}$ is convex; weak convexity means diminishing marginal rate of substitution. It is called *convex* if for any x^i, $x^{i'} \in X^i$ for which $x^i >_i x^{i'}$ and for any real number t for which $0 < t < 1$ it follows that $tx^i + (1-t)x^{i'} >_i x^{i'}$. It is called *strictly convex* if for any two distinct x^i, $x^{i'} \in X^i$ for which $x^i \sim_i x^{i'}$ and for any real number t for which $0 < t < 1$ it follows that $tx^i + (1-t)x^{i'} >_i x^{i'}$. It is called *monotone* if for any x^i, $x^{i'} \in X^i$ for which $x^i > x^{i'}$ it follows that $x^i >_i x^{i'}$; monotonicity means that each commodity is desirable to consumer i. A commodity bundle $x^i \in X^i$ is called a *nonsatiation point* if there exists $x^{i'} \in X^i$ such that $x^{i'} >_i x^i$. A numerical function $u^i \colon X^i \to \mathbf{R}$ *represents* the preference relation $\lesssim_i$ if for any $x^i, x^{i'} \in X^i$, $x^i \lesssim_i x^{i'}$ if and only if $u^i(x^i) \leq u^i(x^{i'})$. A consumer whose preference relation is representable by a numerical function behaves *as though* he were trying to maximize this function. Sometimes such a numerical function is called a *utility function*. Proof of the following theorem (Theorem 0.2.1) can be found in Debreu (1959, Theorem (1), pp. 56–59).

Theorem 0.2.1. *Let X^i be the consumption set of consumer i, and let $\lesssim_i$ be his preference relation. Assume X^i is a connected subset of $\mathbf{R}^l$. Then there exists a continuous numerical function on X^i that represents $\lesssim_i$ if and only if $\lesssim_i$ is complete, transitive, and closed.*

The *initial endowment vector* ω^i is a point in $\mathbf{R}^l$ and is interpreted as the commodity bundle consumer i holds initially. A *pure exchange economy* is now characterized by a list of specified data, $\mathscr{E} := \{X^i, \lesssim_i, \omega^i\}_{i=1}^m$.

Besides the exogenous data $\mathscr{E}$, economists identify the behavioral pattern of the economic agents and the mechanism at the "meeting place" that coordinates their behavior. Consumer behavior in accordance with a given pattern leads to economic outcomes usually characterized by particular values of appropriate endogenous variables. Economic theorists formulate the outcomes in terms of a suitable solution concept and try to understand them by deducing the properties of the solution.

The solution concept with the greatest importance and longest history is the competitive equilibrium: each consumer i observes a price vector $p \in \mathbf{R}^l_+ \setminus \{\mathbf{0}\}$ in the market. His own budget set $\gamma^i(p, p \cdot \omega^i) :=$

0.2. Pure Exchange Economy

$\{\xi^i \in X^i \mid p \cdot \xi^i \le p \cdot \omega^i\}$ is therefore determined, and within this constraint he chooses individualistically the commodity bundle that satisfies him the most. An equilibrium price vector is then determined in the market so that the total demand cannot exceed the total supply. Let $L := \{1, \ldots, l\}$. By the price–wealth homogeneity, one may restrict the price vector domain to $\{p \in \mathbf{R}_+^l \mid \sum_{h=1}^l p_h = 1\}$ that will be, by abuse of notation, denoted by Δ^L throughout this text. Thus the *competitive equilibrium* of a pure exchange economy $\mathscr{E}$ is a pair $((x^{i*})_{i=1}^m, p^*)$ of members of $\prod_{i=1}^m X^i$ and Δ^L such that

(1) x^{i*} is a maximal element of $\{\xi^i \in X^i \mid p^* \cdot \xi^i \le p^* \cdot \omega^i\}$ with respect to $\lesssim_i$ for every i; and
(2) $\sum_{i=1}^m x^{i*} \le \sum_{i=1}^m \omega^i$.

It is a solution concept based on noncooperative behavior of the consumers and on the market mechanism. In Section 4.4, however, its existence problem will be discussed as a particular case of the existence problem of a certain noncooperative solution concept that does not specifically involve the market mechanism. Other solution concepts for $\mathscr{E}$ based on a cooperative behavior will be discussed in Sections 5.5 and 6.4.

Certain dynamic economies can be analyzed within the framework of the above static model $\mathscr{E}$. One characterizes a commodity not only by its physical properties and the place where it is available, but also by the date when it will be available and (in the case of uncertainty about the future) by the elementary event that will be realized. Commodity h, for example, is defined as coffee ice-cream that will be available in Iowa City in 77 days when it is snowing. The above definition of competitive equilibrium allows for this dynamic interpretation, but one should keep in mind that all futures markets (and also all contingent markets in the presence of uncertainty) are assumed to exist.

For the case $l = m = 2$ and $X^i = \mathbf{R}_+^2$ for $i = 1, 2$ the competitive equilibrium $((x^{i*})_{i=1,2}, p^*)$ is illustrated in the *Edgeworth box diagram* (see Figure 0.2.1). A commodity bundle of consumer 1 is measured from the origin $\mathbf{0}^1$ by the $\overrightarrow{\mathbf{0}^1 x_1^1}$ and $\overrightarrow{\mathbf{0}^1 x_2^1}$ axes. A commodity bundle of consumer 2 is measured from the origin $\mathbf{0}^2$ by the $\overrightarrow{\mathbf{0}^2 x_1^2}$ axis and the $\overrightarrow{\mathbf{0}^2 x_2^2}$ axis. The second origin $\mathbf{0}^2$ lies at the point $\omega^1 + \omega^2$ when measured

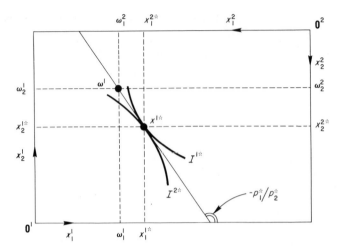

Figure 0.2.1 Edgeworth box diagram and competitive equilibrium.

from the origin 0^1 so it also represents the total supply vector. The curve I^{i*} is the indifference curve $\{x^i \in X^i \mid x^i \sim_i x^{i*}\}$ passing through x^{i*}, $i = 1, 2$. Now a competitive equilibrium is characterized by the following properties: (1) I^{1*} and I^{2*} are tangent at x^{1*} with the slope being $-p_1^*/p_2^*$; and (2) the tangent line passes through ω^1 so that it becomes the budget line.

1

Introduction to Convex Analysis

The basic ingredients of convex analysis are presented in the finite-dimensional setup. A definition of convexity, its immediate consequences, and related concepts are leisurely exposed in Sections 1.1–1.3. The deepest result is that of Theorem 1.4.4 of Section 1.4, in which some (Euclidean) topological concepts for convex sets are characterized in terms of vector space structure. In Section 1.5 two versions of the separation principle are established: the support theorem (Theorem 1.5.3) and the separation theorem (Theorem 1.5.4). Two other versions are also given in Exercises 6 and 7: the Hahn–Banach theorem and the subdifferentiability theorem. These four versions are equivalent in the sense that given any one of them the other three follow immediately. In Section 1.6 the concept of extreme point is introduced, and its elementary existence theorem (Theorem 1.6.1) is provided. Also introduced is a more general concept, the facial space. Artstein's fundamental lemma (Lemma 1.6.4) on facial spaces is presented, and its applications are discussed; in particular, the Shapley–Folkman theorem is proved. Assertions of certain theorems in this chapter are false in the infinite-dimensional context. Indeed, the negation of certain assertions characterizes infiniteness of the dimension of a given vector space. This last issue is discussed in the Appendix to this chapter. For general references pertinent to convex analysis see Fenchel (1951) and Rockaffellar (1970); also see Nikaidô (1968), which contains applications to economics.

1.1. Convex Set

A subset C of $\mathbf{R}^n$ is called *convex* if $[x, y \in C, \alpha \in \mathbf{R}, 0 \leq \alpha \leq 1]$ implies $\alpha x + (1 - \alpha)y \in C$. Given an arbitrary subset S of $\mathbf{R}^n$, one can associate with it naturally a convex set called the convex hull of S. Note that for an indexed family $\{C_i\}_{i \in I}$ of convex subsets of $\mathbf{R}^n$, $\bigcap_{i \in I} C_i$ is a convex set. The *convex hull* of S is the set co $S := \bigcap \{C \mid C$ is a convex subset of $\mathbf{R}^n$, $C \supset S\}$; it is the smallest convex set that contains S. The set co S is now characterized.

Let $(x^i)_{i \in F}$ be a finite set in $\mathbf{R}^n$. A point y in $\mathbf{R}^n$ is called a *convex combination* of $(x^i)_{i \in F}$ if there exists a nonnegative real coefficient α_i for each $i \in F$, with $\sum_{i \in F} \alpha_i = 1$, such that $y = \sum_{i \in F} \alpha_i x^i$. Let S' be the set of all convex combinations of finitely many members of S. By showing that any convex subset of $\mathbf{R}^n$ that contains S also contains S' and that S' itself is convex one can prove

Theorem 1.1.1. *The convex hull of a subset S of $\mathbf{R}^n$ is precisely the set of all convex combinations of finitely many members of S.*

Theorem 1.1.1 holds true for an arbitrary vector space over $\mathbf{R}$. A sharper result in the finite-dimensional context is

Theorem 1.1.2 (Carathéodory). *The convex hull of a subset S of $\mathbf{R}^n$ is precisely the set of all convex combinations of $(n + 1)$ members of S.*

Lemma 1.1.3. *Let $(x^i)_{i \in F}$ be a finite set in $\mathbf{R}^n$ and y be its nonnegative linear combination. Then there exists $F_0 \subset F$, with $\#F_0 \leq n$, such that y is a positive linear combination of $(x^i)_{i \in F_0}$.*

PROOF. Let $y = \sum_{i \in F} \alpha_i x^i$, and assume w.l.o.g. that $\alpha_i > 0$ for all $i \in F$.

Step 1. If $(x^i)_{i \in F}$ is linearly dependent, then $\exists F' \subsetneq F : y$ is a positive linear combination of $(x^i)_{i \in F'}$. Indeed, there exist γ_i, $i \in F$, not all 0, such that $\sum_{i \in F} \gamma_i x^i = \mathbf{0}$. Without loss of generality $\exists i \in F : \gamma_i > 0$ (other-

1.1. Convex Set

wise, multiply by -1). Define $\theta := \min_{i:\gamma_i>0} \alpha_i/\gamma_i$, and consider $y = \sum_{i\in F} (\alpha_i - \theta\gamma_i)x^i$.

Step 2. Repeat the procedure of Step 1 until a linearly independent subset $(x^i)_{i\in F_0}$ is obtained. By linear independence $\#F_0 \le n$. □

PROOF OF THEOREM 1.1.2. Given a finite set $(x^i)_{i\in F}$ in $\mathbf{R}^n$, a vector $y(\in\mathbf{R}^n)$ is a convex combination of $(x^i)_{i\in F}$ iff $(y, 1)$ is a nonnegative linear combination of $((x^i, 1))_{i\in F}$ [with the same coefficients]. So

$$y \in \text{co } S \Leftrightarrow \left[y = \sum_{i\in F} \alpha_i x^i, \#F < \infty, x^i \in S, \alpha_i \in \mathbf{R}_+, \sum_{i\in F} \alpha_i = 1 \right]$$

$$\Leftrightarrow \left[(y, 1) = \sum_{i\in F} \alpha_i(x^i, 1), \#F < \infty, x^i \in S, \alpha_i \in \mathbf{R}_+ \right]$$

$$\Leftrightarrow \left[(y, 1) = \sum_{i\in F_0} \beta_i(x^i, 1), \#F_0 \le n+1, x^i \in S, \beta_i \in \mathbf{R}_+ \right]$$

$$\Leftrightarrow \left[y = \sum_{i\in F_0} \beta_i x^i, \#F_0 \le n+1, x^i \in S, \beta_i \in \mathbf{R}_+, \sum_{i\in F_0} \beta_i = 1 \right]. \quad \Box$$

Corollary 1.1.4. *Let S be a subset of $\mathbf{R}^n$. If S is compact, then so is co S.*

PROOF. Denote by Δ the set $\{\alpha \in \mathbf{R}^{n+1} \mid \alpha \ge 0, \sum_{i=1}^{n+1} \alpha_i = 1\}$, and define the function

$$f: \overbrace{S \times \cdots \times S}^{n+1} \times \Delta \to \mathbf{R}^n$$

by $f(x^1, \ldots, x^{n+1}, \alpha) := \sum_{i=1}^{n+1} \alpha_i x^i$. The function f is continuous on its compact domain $S \times \cdots \times S \times \Delta$, so its image $f(S \times \cdots \times S \times \Delta)$ is compact. But the image is precisely co S by Carathéodory's theorem (Theorem 1.1.2). □

A finite subset $(x^i)_{i\in F}$ of $\mathbf{R}^n$ is called *affinely independent* if

$$\left[\sum_{i\in F} r_i x^i = \mathbf{0}, r_i \in \mathbf{R}, \sum_{i\in F} r_i = 0 \right]$$

implies $r_i = 0$ for each $i \in F$; or, equivalently, if with an arbitrarily chosen $i_0 \in F$ the set $(x^i - x^{i_0})_{i \in F \setminus \{i_0\}}$ is linearly independent. A subset S of $\mathbf{R}^n$ is called a *k-dimensional simplex* if there is an affinely independent set $(x^i)_{i \in F}$, with $\#F = k + 1$, such that $S = \text{co}(x^i)_{i \in F}$; those x^i are called the *vertices* of S. Each point of a simplex is uniquely expressed as a convex combination of the vertices. A simplex is compact and convex. Frequently one can easily prove theorems on a compact, convex set by first establishing the results on a simplex.

1.2. Affine Subspace

A class of convex sets plays a central role in analyzing algebraic and topological properties of (general) convex sets. A subset M of $\mathbf{R}^n$ is called an *affine subspace* if there exist a point m and a subspace W of $\mathbf{R}^n$ such that $M = \{m\} + W$. Given an affine subspace M, such a subspace W is uniquely determined. Indeed,

Lemma 1.2.1. *Let M be an affine subspace of $\mathbf{R}^n$, say $M = \{m\} + W$ for a point m and a subspace W of $\mathbf{R}^n$. Then $m \in M$ and $W = M - M$.*

PROOF. The first of the conclusions is straightforward since $\mathbf{0} \in W$. If $x \in W$, then $2x \in W$. So $m + x, m + 2x \in M$. Therefore $x = (m + 2x) - (m + x) \in M - M$. This proves $W \subset M - M$. Choose any $m^1, m^2 \in M$. There exist $x^i \in W$ such that $m^i = m + x^i$, $i = 1, 2$. Then $m^1 - m^2 = x^1 - x^2 \in W$. Thus $M - M \subset W$. □

The *dimension* of an affine subspace M is defined as the dimension of the unique subspace $(M - M)$. A characterization of an affine subspace is

Theorem 1.2.2. *Let M be a subset of $\mathbf{R}^n$. The set M is an affine subspace iff $[x, y \in M, r \in \mathbf{R}]$ implies $rx + (1 - r)y \in M$.*

1.2. Affine Subspace

PROOF. The statement "only if" is trivial by the definition of an affine subspace. To show the converse, choose any $m \in M$. The identity $x = m + (x - m)$ establishes $M \subset \{m\} + (M - M)$. Applying the given condition to the identity

$$m + (x - y) = \tfrac{1}{2}(2m + (1 - 2)y) + (1 - \tfrac{1}{2})(2x + (1 - 2)y),$$

one establishes the other inclusion $\{m\} + (M - M) \subset M$. It suffices to show that $(M - M)$ is a subspace of $\mathbf{R}^n$. Take any $r \in \mathbf{R}$, and any $x, y \in M$. Then $r(x - y) = (rx + (1 - r)y) - y \in M - M$. This proves $\mathbf{R} \cdot (M - M) \subset M - M$. Take any $x^i, y^i \in M$, $i = 1, 2$. Then

$$(x^1 - y^1) + (x^2 - y^2) = 2((\tfrac{1}{2}x^1 + \tfrac{1}{2}x^2)$$

$$- (\tfrac{1}{2}y^1 + \tfrac{1}{2}y^2)) \in \mathbf{R} \cdot (M - M) \subset M - M.$$

This proves $(M - M) + (M - M) \subset M - M$. □

Note that for an indexed family $\{M_i\}_{i \in I}$ of affine subspaces of $\mathbf{R}^n$ the intersection $\bigcap_{i \in I} M_i$ is also an affine subspace; indeed, if $M_i = \{m^i\} + W_i$, then for any $m \in \bigcap_{i \in I} M_i$ it follows that $\bigcap_{i \in I} M_i = \{m\} + \bigcap_{i \in I} W_i$. Let S be an arbitrary subset of $\mathbf{R}^n$. The *affine hull* of S is defined as the set aff $S := \bigcap \{M \mid M \text{ is an affine subspace of } \mathbf{R}^n, M \supset S\}$; it is the smallest affine subspace that contains S. Denote by span S the subspace generated by S (i.e., the smallest subspace of $\mathbf{R}^n$ that contains S).

Theorem 1.2.3. *Let S be a subset of $\mathbf{R}^n$. Then for any $m \in$ aff S (in particular, for any $m \in S$), aff $S = \{m\} +$ span$(S - S)$.*

PROOF. Choose any $m \in S$. Since aff $S -$ aff S is a subspace that contains $S - S$, $\{m\} +$ span$(S - S) \subset \{m\} +$ (aff $S -$ aff S) = aff S. To show the other inclusion it suffices to show that the affine subspace $\{m\} +$ span $(S - S)$ contains S. But this is straightforward by the identity $x = m + (x - m)$. Thus aff $S = \{m\} +$ span$(S - S)$. By Lemma 1.2.1, span$(S - S) =$ aff $S -$ aff S, one can now replace m by any member of aff S. □

Another characterization of the affine hull is easily obtained from either one of the preceding two theorems:

Corollary 1.2.4. *Let S be a subset of $\mathbf{R}^n$. Then x is a member of aff S iff it is a linear combination of finitely many members of S such that the coefficients are summed to 1.*

Note that all statements made in this section hold true for an arbitrary vector space over $\mathbf{R}$.

1.3. Hyperplane

An $(n-1)$-dimensional affine subspace in $\mathbf{R}^n$ is called a *hyperplane*.

Theorem 1.3.1. *Let M be a subset of $\mathbf{R}^n$. The set M is a hyperplane iff there exist $h \in \mathbf{R}^n \setminus \{0\}$ and $r \in \mathbf{R}$ such that $M = \{x \in \mathbf{R}^n \mid h \cdot x = r\}$.*

PROOF. *Necessity.* Put $M = \{m\} + W$, where W is an $(n-1)$-dimensional subspace. Then there exists $\bar{x} \in \mathbf{R}^n \setminus W$, and clearly $\mathbf{R}^n = \mathrm{span}(W \cup \{\bar{x}\})$. Define a linear map $h: \mathbf{R}^n \to \mathbf{R}$ (that is represented by an n-dimensional vector, given the standard basis) by: $h(x) = 0$ if $x \in W$, and $h(\bar{x}) = 1$. Put $r := h(m)$. It is straightforward to check that $M = \{x \in \mathbf{R}^n \mid h \cdot x = r\}$ since the kernel of h is precisely W.

Sufficiency. Choose any $m \in M$, and just observe that $M = \{m\} + \ker h$. □

The nonzero vector h in the preceding theorem is called a *normal vector* of M.

A hyperplane can also be defined for an arbitrary vector space V over $\mathbf{R}$: Let $\mathscr{M}$ be the family of all proper affine subspaces of V. The family $\mathscr{M}$, ordered by the inclusion $\subset$, is inductively ordered and hence it has a maximal element. A *hyperplane* in V is a maximal element in $(\mathscr{M}, \subset)$. An analogy of the above characterization of hyperplanes in terms of the normal vectors also holds true, where h is a nonzero member of the algebraic dual V' of V.

1.4. Algebraic Interior, Algebraic Relative Interior

Let x, y be points in $\mathbf{R}^n$. Notation for segments joining x and y is given by

$$[x, y] := \{\alpha x + (1 - \alpha)y \mid \alpha \in \mathbf{R}, 0 \le \alpha \le 1\},$$
$$]x, y[:= \{\alpha x + (1 - \alpha)y \mid \alpha \in \mathbf{R}, 0 < \alpha < 1\},$$
$$[x, y[:= \{\alpha x + (1 - \alpha)y \mid \alpha \in \mathbf{R}, 0 < \alpha \le 1\}.$$

The set $]x, y]$ is similarly defined. Note that if $x = y$, all four concepts are identical to $\{x\}$.

Let S, T be subsets of $\mathbf{R}^n$. The algebraic interior of S relative to T is the set

$$\mathrm{cor}_T(S) := \{x \in S \mid \forall y \in T \setminus \{x\} : \exists z \in]x, y] : [x, z] \subset S\}.$$

The *algebraic interior* of S (*algebraic relative interior* of S, resp.) is the set $\mathrm{cor}_{\mathbf{R}^n}(S)$ (the set $\mathrm{cor}_{\mathrm{aff}\,S}(S)$, resp.), and is denoted by cor S (by icr S, resp.). The notation cor S (icr S, resp.) stands for the term "core" ("intrinsic core," resp.). This terminology will not be adopted in the present text, however, since core is now a standard term for one of the central concepts in game theory; see Chapter 5.

A nonempty simplex always has a nonempty algebraic relative interior; indeed, if $(x^i)_{i \in F}$ is an affinely independent, finite subset of $\mathbf{R}^n$, then

$$\mathrm{icr}\, \mathrm{co}(x^i)_{i \in F} = \{\sum_{i \in F} \alpha_i x^i \mid \forall i \in F : \alpha_i > 0, \sum_{i \in F} \alpha_i = 1\}$$

since aff $\mathrm{co}(x^i)_{i \in F}$ is precisely the set

$$\{\sum_{i \in F} r_i x^i \mid \forall i \in F : r_i \in \mathbf{R}, \sum_{i \in F} r_i = 1\}.$$

A more general fact is

Theorem 1.4.1. *A nonempty, convex subset of $\mathbf{R}^n$ has a nonempty algebraic relative interior.*

PROOF. Let C be a nonempty, convex subset of $\mathbf{R}^n$. Choose an affinely independent, finite subset $(x^i)_{i \in F}$ of C of highest cardinality (which is possible because no set of cardinality greater than or equal to $(n + 2)$ can be affinely independent in $\mathbf{R}^n$). Observe that $\mathrm{co}(x^i)_{i \in F} \subset C \subset \mathrm{aff}(x^i)_{i \in F}$. (The second inclusion follows from the fact that given any $x \in C$ the set $\{x, (x^i)_{i \in F}\}$ cannot be affinely independent, and so w.l.o.g., there exist r_i such that $(-1)x + \sum_{i \in F} r_i x^i = \mathbf{0}$ and $(-1) + \sum_{i \in F} r_i = 0$.) Therefore $\mathrm{aff}\, C = \mathrm{aff}(x^i)_{i \in F}$, so $\emptyset \neq \mathrm{icr}\, \mathrm{co}(x^i)_{i \in F} \subset \mathrm{icr}\, C$. □

The following lemma provides a convenient characterization of the algebraic relative interior of a convex set.

Lemma 1.4.2. *Let C be a nonempty, convex subset of $\mathbf{R}^n$. Choose any $\bar{x} \in \mathrm{icr}\, C$. Then $\mathrm{icr}\, C = \bigcup \{[\bar{x}, x[\,|\, x \in C\}$.*

PROOF. It is trivial that the left-hand side is contained in the right-hand side. To show the other inclusion choose $x \in C$ and $y \in [\bar{x}, x[$. One has to show that $y \in \mathrm{icr}\, C$ (see Figure 1.4.1). Take any $z \in (\mathrm{aff}\, C) \setminus \{y\}$. Since $\bar{x} \in \mathrm{icr}\, C$, there exists $v \in \,]\bar{x}, \bar{x} + z - y[\,\cap\, C$. Let v' be the point of intersection of $[v, x]$ and $]y, z[$. Then $[y, v'] \subset C$. □

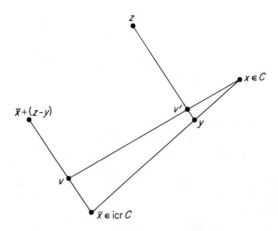

Figure 1.4.1

1.4. Algebraic Interior, Algebraic Relative Interior

Let C be a nonempty, convex subset of $\mathbf{R}^n$. It is clear that cor $C = \emptyset$ if aff $C \neq \mathbf{R}^n$ and that cor $C = $ icr $C \neq \emptyset$ if aff $C = \mathbf{R}^n$.

Let x and S be a point and a subset of $\mathbf{R}^n$, respectively. The point x is called *linearly accessible from* S if there exists a point $v \in S$, distinct from x, such that $]v, x[\subset S$. Denote by lina S the set of all points that are linearly accessible from S, and define lin $S := S \cup $ lina S.

Denote by $\overset{\circ}{S}$ (by $\bar{S}$, resp.) the interior (the closure, resp.) of a set S in $\mathbf{R}^n$. The *relative interior* of S is the interior of S in aff S. In general

$$\overset{\circ}{S} \subset \text{cor } S \subset \text{icr } S \subset S \subset \text{lin } S \subset \bar{S}.$$

For each inclusion, one can construct an example to make it strict (see Figures 1.4.2 and 1.4.3). The situation becomes pleasant when S is convex; the topological concepts $\overset{\circ}{S}$ and $\bar{S}$ are then characterized as the algebraic concepts cor S and lin S.

Lemma 1.4.3. *Let C be a convex subset of $\mathbf{R}^n$ with a nonempty interior, and let r be a real number with $0 \leq r < 1$. Then $r\bar{C} + (1 - r)\overset{\circ}{C} \subset \overset{\circ}{C}$.*

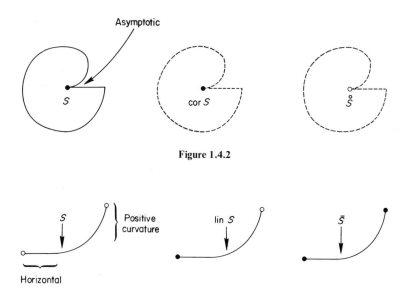

Figure 1.4.2

Figure 1.4.3

PROOF. Since the set $C' := r\bar{C} + (1-r)\mathring{C}$ is open, it suffices to show that $C' \subset C$. Choose any $\bar{x} \in \mathring{C}$. Then $-(1-r)(\mathring{C} - \{\bar{x}\})$ is a neighborhood of $\mathbf{0}$. Since $\overline{rC}$ is the set of all cluster points of rC,

$$r\bar{C} = \overline{rC} \subset rC + (1-r)(\mathring{C} - \{\bar{x}\})$$
$$= rC + (1-r)\mathring{C} - (1-r)\{\bar{x}\}$$
$$\subset C - (1-r)\{\bar{x}\}.$$

Thus $r\bar{C} + (1-r)\{\bar{x}\} \subset C$ for all $\bar{x} \in \mathring{C}$. □

Theorem 1.4.4. *Let C be a convex subset of $\mathbf{R}^n$. Then $\mathring{C} = \operatorname{cor} C$, [the relative interior of C] = icr C, and $\bar{C} = \operatorname{lin} C$.*

PROOF. If aff $C \neq \mathbf{R}^n$, then $\mathring{C} = \operatorname{cor} C = \emptyset$. If aff $C = \mathbf{R}^n$, then, as in the proof of Theorem 1.4.1, one can choose an affinely independent subset $(x^i)_{i \in F}$ of C, with $\#F = n + 1$. Clearly,

$$\emptyset \neq \Big\{ \sum_{i \in F} \alpha_i x^i \,\Big|\, \forall i \in F : \alpha_i > 0, \sum_{i \in F} \alpha_i = 1 \Big\} \subset \mathring{C}.$$

Fix any $\bar{x} \in \mathring{C}$. If $x \in \operatorname{cor} C$, then there exists $y \in C$ such that $x \in]y, \bar{x}]$, i.e., $x \in rC + (1-r)\mathring{C}$ for some $r : 0 \leq r < 1$. By Lemma 1.4.3, $x \in \mathring{C}$. Thus cor $C \subset C$. The same argument establishes the second result. To show the last result, one may assume w.l.o.g. that aff $C = \mathbf{R}^n$ (hence $\mathring{C} \neq \emptyset$) because $\bar{C} \subset \operatorname{aff} C$. Fix any $\bar{x} \in \mathring{C}$. If $x \in \bar{C}$, then by Lemma 1.4.3, $rx + (1-r)\bar{x} \in \mathring{C}$ for all $r : 0 \leq r < 1$. So $x \in \operatorname{lin} C$. □

Remark that the closure of a convex set C in $\mathbf{R}^n$ is convex because the function $f : \mathbf{R}^n \times \mathbf{R}^n \times [0, 1] \to \mathbf{R}^n$, $(x, y, \alpha) \mapsto \alpha x + (1 - \alpha) y$ is continuous, so

$$f(\bar{C} \times \bar{C} \times [0, 1]) = f(\overline{C \times C \times [0, 1]}) \subset \overline{f(C \times C \times [0, 1])} = \bar{C}.$$

The preceding theorem shows that the interior of a convex set C in $\mathbf{R}^n$ is also convex because it is identical to cor C, which in turn is convex by Lemma 1.4.2. Another consequence of the theorem is

Corollary 1.4.5. *Let C be a convex subset of $\mathbf{R}^n$. Then the closure in*

1.5. Separation Principle

$\mathbf{R}^n$ *of the relative interior of C is the closure of C in $\mathbf{R}^n$, and the relative interior of $\bar{C}$ is the relative interior of C.*

PROOF. Again, one may assume w.l.o.g. that $C \neq \emptyset$ because icr $C \neq \emptyset$ and $\bar{C} \subset$ aff C. If $x \in \bar{C}$, then the same argument as in the proof of the last statement of the preceding theorem shows that $x \in $ lina $\overset{\circ}{C}$ so $x \in \overline{\overset{\circ}{C}}$. If $x \in \overset{\circ}{\bar{C}}$, then $x \in \text{cor}(\bar{C})$ since $\bar{C}$ is convex. Choose $\bar{x} \in \overset{\circ}{C}$. Then there exists $y \in \bar{C}$ such that $x \in \,]y, \bar{x}]$. By Lemma 1.4.3, $x \in \overset{\circ}{C}$. □

The assertion of Corollary 1.4.5 does not hold true in general for a nonconvex set C; consider, for example, the set of the rational numbers in the real line.

Many of the results in this section fail to hold in the infinite-dimensional context. In particular a vector space V over $\mathbf{R}$ is infinite-dimensional iff there exists a nonempty, convex subset C of V with icr $C = \emptyset$; the preceding theorem needs an additional assumption of $\overset{\circ}{\bar{C}} \neq \emptyset$. Note also that a vector space V over $\mathbf{R}$ is infinite-dimensional iff there exists a convex, proper subset C of V such that lin $C = V$.

1.5. Separation Principle

There are many useful versions of one basic principle in convex analysis. The principle is henceforth called the *separation principle* because one of the versions says that given two disjoint, convex sets one of them can be located in a half-space determined by a hyperplane and the other set in the other half-space. Once any version is established, the other versions follow quite easily. Probably the best method for establishing one version is to exploit properties of the so-called complementary convex sets, using only the algebraic structure; such a method was developed by Kakutani (1937) and Stone (1946). An easier method, due to von Neumann and Morgenstern (1947), that uses both the algebraic structure and the topological structure of $\mathbf{R}^n$ is presented here.

Theorem 1.5.1. *Let C be a closed, convex subset of $\mathbf{R}^n$, and let v be a point in $\mathbf{R}^n \backslash C$. Then there exist $h \in \mathbf{R}^n \backslash \{0\}$ and $r \in \mathbf{R}$ such that $h \cdot v < r$ and $h \cdot x \geq r$ for all $x \in C$.*

PROOF. By the closedness of C there exists $w \in C$ for which $\|w - v\| \leq \|x - v\|$ for every $x \in C$. Set $h := w - v$, and $r := h \cdot w$. Clearly, $h \neq 0$ and $h \cdot v < r$. Choose any $x \in C$. It will be shown that $h \cdot x \geq r$. For each $\alpha \in \mathbf{R} : 0 < \alpha \leq 1$ define $x(\alpha) := (1 - \alpha)w + \alpha x$. Then $x(\alpha) \in C$ by the convexity assumption. Therefore

$$\|w - v\|^2 \leq \|x(\alpha) - v\|^2$$
$$= (1 - \alpha)^2 \|w - v\|^2$$
$$+ 2\alpha(1 - \alpha)(w - v) \cdot (x - v) + \alpha^2 \|x - v\|^2.$$

By rearranging this and dividing both sides by α,

$$(\alpha - 2)\|w - v\|^2 + 2(1 - \alpha)(w - v) \cdot (x - v) + \alpha \|x - v\|^2 \geq 0.$$

Let $\alpha \to 0$. Then in the limit

$$-\|w - v\|^2 + (w - v) \cdot (x - v) \geq 0,$$

i.e., $h \cdot x \geq r$. □

Corollary 1.5.2. *Let C be a convex subset of $\mathbf{R}^n$, and let v be a point in the boundary of C. Then there exists $h \in \mathbf{R}^n \backslash \{0\}$ such that $h \cdot x \geq h \cdot v$ for all $x \in C$.*

PROOF. If $\overset{\circ}{C} = \varnothing$, then aff $C \neq \mathbf{R}^n$ so a normal vector of any hyperplane that contains aff C is the required h. Assume, therefore, $\overset{\circ}{C} \neq \varnothing$; then $\overset{\circ}{C} = $ icr C. By Corollary 1.4.5 v is also in the boundary of $\bar{C}$. There exists a sequence $\{v^k\}_k$ in $\mathbf{R}^n \backslash \bar{C}$ that converges to v. For each k apply Theorem 1.5.1; there exists $h^k \in \mathbf{R}^n \backslash \{0\}$ such that $h^k \cdot v^k < h^k \cdot x$ for all $x \in \bar{C}$. One may assume w.l.o.g. that $\|h^k\| = 1$ for every k. Then the sequence $\{h^k\}_k$ has a convergent subsequence, say $\{h^{k_p}\}_p$ with $h^{k_p} \to h$ as $p \to \infty$. It is straightforward to check that the limit h satisfies all the required properties. □

1.5. Separation Principle

Theorem 1.5.3. **(Support Theorem).** *Let C be a convex subset of $\mathbf{R}^n$, and let v be a point in $\mathbf{R}^n\backslash\operatorname{icr} C$. Then there exists $h \in \mathbf{R}^n\backslash\{0\}$ such that $h \cdot x > h \cdot v$ for all $x \in \operatorname{icr} C$ (see Figure 1.5.1).*

PROOF. *Case 1.* Assume that $\overset{\circ}{C} \neq \varnothing$. If $v \notin \bar{C}$, then Theorem 1.5.1 gives the required vector h. If $v \in \bar{C}$ apply the preceding corollary; there exists $h \in \mathbf{R}^n\backslash\{0\}$ such that $h \cdot x \geq h \cdot v$ for all $x \in C$. It suffices to show that the inequality is strict for each $x \in \overset{\circ}{C}$. Suppose the contrary: $\exists \bar{x} \in \overset{\circ}{C}; h \cdot \bar{x} = h \cdot v$. Then for $\epsilon > 0$ small enough $\bar{x} - \epsilon h \in \overset{\circ}{C}$, and yet $h \cdot (\bar{x} - \epsilon h) < h \cdot v$—a contradiction.

Case 2. Assume that $\overset{\circ}{C} = \varnothing$. If $v \notin \operatorname{aff} C$, then Theorem 1.5.1 is applicable to the point v and the closed, convex set aff C. Therefore, suppose that $v \in \operatorname{aff} C$. Choose any $m \in \operatorname{aff} C$. Then the convex set $C - \{m\}$ has a nonempty interior in the subspace (aff C) $- \{m\}$, and this interior does not contain $v - m$. By Case 1 there exists a linear map $h: ((\operatorname{aff} C) - \{m\}) \to \mathbf{R}$ such that $h(x - m) > h(v - m)$, hence $h(x) > h(v)$, for all $x - m$ in the interior of $C - \{m\}$ in (aff C) $- \{m\}$, i.e., for all x in icr C. Extend h linearly to $\mathbf{R}^n$. □

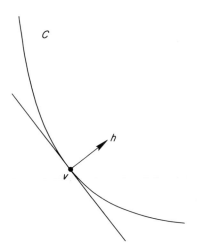

Figure 1.5.1 Support theorem.

Let C be a convex subset of $\mathbf{R}^n$. A hyperplane H is said to *support* C if C lies in one of the two half-spaces determined by H and $C \cap H \neq \emptyset$. Given a supporting hyperplane H of C that does not completely contain C, the members of $C \cap H$ are called *proper support points* of C. The preceding support theorem implies: the set of proper support points of C is precisely $C \backslash \mathrm{icr}\, C$.

Theorem 1.5.4 (Separation Theorem). *Let C, D be two disjoint, nonempty, convex subsets of $\mathbf{R}^n$. Then there exist $h \in \mathbf{R}^n \backslash \{0\}$ and $r \in \mathbf{R}$ such that $h \cdot x \geq r$ for every $x \in C$ and $h \cdot y \leq r$ for every $y \in D$ (see Figure 1.5.2).*

PROOF. The set $C - D$ is convex, and $0 \notin C - D$. So, by the support theorem, there exists $h \in \mathbf{R}^n \backslash \{0\}$ such that $h \cdot (x - y) \geq h \cdot 0$, hence $h \cdot x \geq h \cdot y$, for all $x \in C$ and $y \in D$. Since $\{h \cdot x \mid x \in C\}$ is bounded from below by $h \cdot y$ for an arbitrarily chosen $y \in D$, $r := \inf\{h \cdot x \mid x \in C\} \in \mathbf{R}$. It is straightforward to verify that $h \cdot y \leq r$ for all $y \in D$. □

Let C, D be convex subsets of $\mathbf{R}^n$. A hyperplane $H := \{x \in \mathbf{R}^n \mid h \cdot x = r\}$ is said to *separate* C and D if $h \cdot x \geq r$ for every $x \in C$ and $h \cdot y \leq r$ for

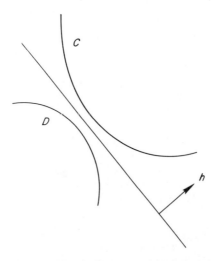

Figure 1.5.2 Separation theorem.

1.6. Extreme Points

every $y \in D$. It is said to *strongly separate* C and D if there exists a positive real number ϵ such that $h \cdot x \geq r + \epsilon$ for every $x \in C$ and $h \cdot y \leq r - \epsilon$ for every $y \in D$.

Theorem 1.5.5 (**Strong Separation Theorem**). *Let C, D be two disjoint, nonempty convex subsets of $\mathbf{R}^n$. The sets C and D are strongly separated by a hyperplane iff there exists a neighborhood U of $\mathbf{0}$ in $\mathbf{R}^n$ such that $(C + U) \cap D = \emptyset$.*

PROOF. *Necessity.* Let $h \in \mathbf{R}^n \backslash \{\mathbf{0}\}, r \in \mathbf{R}, \epsilon > 0$ be such that $h(C) \geq r + \epsilon > r - \epsilon \geq h(D)$. Define $U := \{x \in \mathbf{R}^n \mid -\epsilon < h(x) < \epsilon\}$.

Sufficiency. Without loss of generality the $\mathbf{0}$-neighborhood U is convex and $U = -U$. So $C + U$ and D are two disjoint, convex sets. There exists $h \in \mathbf{R}^n \backslash \{\mathbf{0}\}$ such that $h(x + u - y) \geq 0$ for all $x \in C, u \in U$, and $y \in D$. Define $\epsilon > 0$ so that $2\epsilon = h(\bar{u})$ for some $\bar{u} \in U$, and set $r := \inf\{h(x) - \epsilon \mid x \in C\}$. Take any $y \in D$. Since $h(y) \leq h(x) + h(u)$ for all $x \in C, u \in U$, and since $-\bar{u} \in U$, it follows that $h(y) \leq (r + \epsilon) - 2\epsilon$. □

The present proof of the support theorem depends heavily on the results of Section 1.4, which in turn are based on the fact that icr $C \neq \emptyset$ for any nonempty, convex set C with a finite-dimensional affine hull. This is the reason why, in the infinite-dimensional context, nonemptiness of cor C is assumed in the analogous theorems of the separation principle.

1.6. Extreme Points

Let S be a subset of $\mathbf{R}^n$. A member v of S is called an *extreme point* of S if $v \in]x, y[\subset S$ implies $x = y = v$. For example, the vertices of a simplex S are precisely the extreme points of S.

Theorem 1.6.1. *A nonempty, compact subset of $\mathbf{R}^n$ has an extreme point.*

PROOF. Let S be a nonempty, compact subset of $\mathbf{R}^n$. Given an affine subspace M of $\mathbf{R}^n$, denote by dim M the dimension of M.

Step 1. The theorem is established first with the additional assumption that the set S is convex. The proof proceeds by induction on dim aff S. The assertion is trivial if dim aff $S = 0$. Assume that the assertion is true if dim aff $S \leq k - 1$. The assertion is now proved for the case dim aff $S = k$. Choose $v \in S \backslash \text{icr } S$. This is possible because S is compact (apply Lemma 1.4.2). Then by the support theorem (Theorem 1.5.3), there exists a supporting hyperplane H in $\mathbf{R}^n$ such that $v \in H \cap S$ and such that $H \cap (\text{icr } S) = \emptyset$. Notice that dim aff$(H \cap S) \leq k - 1$, so by the induction hypothesis, the nonempty, compact, convex set $H \cap S$ has an extreme point, say w. It suffices to show that w is an extreme point of S. Let $w \in \,]x, y[\, \subset S$. If both x and y are in H, then $w \in \,]x, y[\, \subset H \cap S$, so $w = x = y$. If one of them is not in H, then the hyperplane H strongly separates x and y (i.e., for a normal vector h of H, $h \cdot x < h \cdot w < h \cdot y$). This contradicts the definition of the supporting hyperplane H since both x and y are members of S.

Step 2. The convexity assumption on S is dropped here. By Corollary 1.1.4, co S is a nonempty, compact, convex subset of $\mathbf{R}^n$. By Step 1, co S has an extreme point. Any extreme point w of co S is an extreme point of S. Indeed, the point w may be written as a convex combination of points in S: $w = \sum_{i \in F} \alpha_i x^i$. If there exists $i \in F$ with $\alpha_i = 1$, then the required result is trivial. Assume therefore that there exists i_0 with $0 < \alpha_{i_0} < 1$. Then

$$w = \alpha_{i_0} x^{i_0} + (1 - \alpha_{i_0}) \sum_{i \in F \backslash \{i_0\}} \left(\frac{\alpha_i}{1 - \alpha_{i_0}} \right) x^i.$$

Since w is an extreme point of co S, it follows that $w = x^{i_0} \in S$. □

Theorem 1.6.2 (Krein–Milman). *Let S be a compact subset of $\mathbf{R}^n$, and let S' be the set of the extreme points of S. Then $\overline{(\text{co } S')} = \overline{(\text{co } S)}$.*

PROOF. It is straightforward that $\overline{(\text{co } S')} \subset \overline{(\text{co } S)}$. To show the other inclusion it suffices to prove that $S \subset \overline{(\text{co } S')}$. Suppose there exists $v \in S \backslash \overline{(\text{co } S')}$. Then by Theorem 1.5.1 there exists $h \in \mathbf{R}^n \backslash \{0\}$ such that $h \cdot v > h \cdot x$ for all $x \in \overline{(\text{co } S')}$. Since S is compact, there exists $v' \in S$ such that $h \cdot v' \geq h \cdot y$ for all $y \in S$; clearly $h \cdot v' \geq h \cdot v$. Denote by H the

1.6. Extreme Points

hyperplane $\{y \in \mathbf{R}^n \mid h \cdot y = h \cdot v'\}$. The nonempty, compact set $H \cap S$ has an extreme point, say w. By the same argument as in Step 1 of the proof of Theorem 1.6.1, the point w is also an extreme point of S—a contradiction to $h \cdot v > h \cdot \overline{(\text{co } S')}$. □

Theorems 1.6.1 and 1.6.2 hold true even for a nonempty, compact subset of a Hausdorff locally convex topological vector space. The proof proceeds by applying Zorn's lemma and a separation theorem directly to a (possibly nonconvex) compact set S. Actually, one can obtain a sharper result in the finite-dimensional context:

Corollary 1.6.3. *Let S be a compact subset of $\mathbf{R}^n$, and let S' be the set of the extreme points of S. Then*

$$\text{co } S' = \overline{(\text{co } S')} = \overline{(\text{co } S)} = \text{co } S.$$

PROOF. The second and third equalities have already been established (Theorems 1.6.2 and 1.1.4). The first equality will be proved by induction on dim aff S. The assertion is trivial if dim aff $S = 0$. Assume that the assertion is true if dim aff $S \leq k - 1$. The assertion is now proved for the case dim aff $S = k$. Choose any point $x \in \text{co } S$. If $x \in \text{icr co } S$, then $x \in \text{co } S'$ by the second equality of the present theorem and by Corollary 1.4.5. Assume, therefore, that $x \notin \text{icr co } S$. By the support theorem (Theorem 1.5.3), there exists a hyperplane H in $\mathbf{R}^n$ such that $x \in H \cap \text{co } S$ and such that $H \cap (\text{icr co } S) = \emptyset$. The set $T := H \cap \text{co } S$ is compact and convex, and dim aff $T \leq k - 1$. Denote by T' the set of extreme points of T. By the induction hypothesis co $T' = T$. Any member of T' is an extreme point of co S (by the same argument as that in Step 1 of the proof of Theorem 1.6.1), hence is an extreme point of S (by the same argument as that in Step 2 of the proof of Theorem 1.6.1), i.e., a member of S'. Thus the point x ($x \in T$) is a convex combination of members of S'. □

Extreme points play central roles in mathematical optimization problems; see, e.g., Rockafellar (1970). Optimization problems will not, however, be discussed in this section. Instead, the concept of extreme point will be used to establish the Shapley–Folkman theorem: The vector sum of "many" subsets in $\mathbf{R}^n$ is "almost" convex. The theorem was first perceived by Ross Starr, and indeed appeared in Starr (1969, Appendix 2). The geometric insight underlying this theorem was made more lucid by subsequent writers, e.g., Howe (1979). The present text follows the work of Artstein (1980).

Let C be a convex subset of $\mathbf{R}^n$, and let x be a member of C. The *facial space* of x relative to C is the set

$$W_C(x) := \{y \in \mathbf{R}^n \mid \exists t > 0 : [x - ty, x + ty] \subset C\};$$

it is a subspace of $\mathbf{R}^n$, and its dimension is denoted by dim $W_C(x)$. An alternative definition: $W_C(x)$ is the maximal subspace T of $\mathbf{R}^n$ for which $x \in \text{cor}_{T + \{x\}}(C)$. Clearly, x is an extreme point of C iff dim $W_C(x) = 0$. A function $f : \mathbf{R}^n \to \mathbf{R}^m$ is called *affine* if there exists $y^0 \in \mathbf{R}^m$ such that the map $\mathbf{R}^n \to \mathbf{R}^m$, $x \mapsto f(x) - y^0$, is linear; note that $y^0 = f(\mathbf{0})$.

Lemma 1.6.4. *Let C be a compact, convex subset of $\mathbf{R}^n$, and let $f : \mathbf{R}^n \to \mathbf{R}^m$ be an affine function. Then for each $y \in f(C)$, there exists $x \in C$ such that $y = f(x)$ and* dim $W_C(x) \leq$ dim $W_{f(C)}(y)$.

PROOF. It is easy to see that the set $C \cap f^{-1}(y)$ is closed and hence compact, so it has an extreme point x. Set $f^0(\cdot) := f(\cdot) - f(\mathbf{0})$. First it will be shown that the linear map f^0 maps $W_C(x)$ into $W_{f(C)}(y)$. Indeed, for each $z \in W_C(x)$ there exists $t > 0$ such that $x + v \in C$ for all $v \in [-tz, tz]$. So $y + f^0(v) = f(x + v) \in f(C)$ for such v. Therefore $f^0(z) \in W_{f(C)}(y)$. The proof of Lemma 1.6.4 will be completed by showing that the map $f^0|_{W_C(x)}$ is one to one. If $f^0(z) = \mathbf{0}$, then for any $t > 0$, $f^0(\pm tz) = \mathbf{0}$ so $f(x \pm tz) = y$. In particular, if the kernel of f^0 contains a nonzero vector $z \in W_C(x)$, then for all $t > 0$ sufficiently small, it follows that $x \pm tz \in C \cap f^{-1}(y)$; this contradicts the choice of x as an extreme point of $C \cap f^{-1}(y)$. □

Given any finitely many nonempty subsets $\{S_i\}_{i \in F}$ of $\mathbf{R}^n$, it is straightforward that co $\sum_{i \in F} S_i = \sum_{i \in F}$ co S_i.

Theorem 1.6.5 (Shapley–Folkman). *Let $\{S_i\}_{i \in F}$ be finitely many nonempty subsets of $\mathbf{R}^n$. Each member y of* co $\sum_{i \in F} S_i$ *has a representation*

1.6. Extreme Points

$y = \sum_{i \in F} x^i$ such that $x^i \in \text{co } S_i$ for all $i \in F$ and such that $x^i \notin S_i$ for at most n of the indices i.

PROOF. One may assume w.l.o.g. that each S_i is compact; otherwise choose $y^i \in \text{co } S_i$ so that $y = \sum_{i \in F} y^i$, and consider a finite subset of S_i to whose convex hull the point y^i belongs. Define an affine map $f: (\mathbf{R}^n)^F \to \mathbf{R}^n$ by $f((x^i)_{i \in F}) := \sum_{i \in F} x^i$ and a nonempty, compact, convex subset C of $(\mathbf{R}^n)^F$ by $C := \prod_{i \in F} \text{co } S_i$. By Lemma 1.6.4, there exists a point $(x^i)_{i \in F} \in C$ such that $y = \sum_{i \in F} x^i$ and $\dim W_C((x^i)_{i \in F}) \le \dim W_{f(C)}(y)$. Since $f(C) \subset \mathbf{R}^n$, $\dim W_{f(C)}(y) \le n$. Thus

$$\sum_{i \in F} \dim W_{\text{co } S_i}(x^i) = \dim W_C((x^i)_{i \in F}) \le n.$$

For all but at most n of the i, it follows that $\dim W_{\text{co } S_i}(x^i) = 0$, i.e., x^i is an extreme point of co S_i. But an extreme point of co S_i has to be a member of S_i. □

There is an obvious analogy between the Shapley–Folkman theorem (Theorem 1.6.5) and Richter's (1963) version of the Lyapunov theorem (Lyapunov, 1940), which says that the integral of a set-valued map with respect to an atomless measure is convex. Indeed, Artstein (1980) also applied his fundamental lemma (Lemma 1.6.4) to prove various versions of the Lyapunov theorem. The use of extreme points for proving the Lyapunov theorem was preceded, however, by Lindenstrauss (1966).

Remark that Step 1 of the proof of Theorem 1.6.1 does not use Corollary 1.1.4 (this corollary is used only in Step 2 in which the given set S may not be convex). Remark also that in the proof of Lemma 1.6.4 one can easily establish convexity of the set $C \cap f^{-1}(y)$. Thus Artstein's fundamental lemma (Lemma 1.6.4) is not based on Corollary 1.1.4, nor on Carathéodory's theorem (Theorem 1.1.2) by which Corollary 1.1.4 is established. Artstein (1980) uses Lemma 1.6.4 to give an alternative proof of Carathéodory's theorem:

ALTERNATIVE PROOF OF CARATHÉODORY'S THEOREM (THEOREM 1.1.2). Choose any $y \in \text{co } S$. Point y is represented as a convex combination of finitely many points of $S: y = \sum_{i \in F} \alpha_i x^i$. Define

$$C := \{\lambda \in \mathbf{R}^F | \lambda \ge \mathbf{0}, \sum_{i \in F} \lambda_i = 1\},$$

a ($\#F - 1$)-dimensional simplex. Define also an affine map $f: \mathbf{R}^F \to \mathbf{R}^n$ by $f(\lambda) := \sum_{i \in F} \lambda_i x^i$. Then $y \in f(C)$, so by the fundamental lemma (Lemma 1.6.4), there exists $\beta \in C$ such that $y = \sum_{i \in F} \beta_i x^i$ and $\dim W_C(\beta) \leq \dim W_{f(C)}(y) \leq n$. It suffices to check that $[\dim W_C(\beta) \leq n]$ implies $[\beta_i = 0$ for all but at most $(n + 1)$ of the $i]$. Set

$$F_0 := \{i \in F \mid \beta_i > 0\}$$

and

$$Z := \{\zeta \in \mathbf{R}^F \mid \sum_{i \in F_0} \zeta_i = 0, \zeta_i = 0 \text{ if } i \notin F_0\}.$$

Then, for each $\zeta \in \mathbf{R}^F$, $[\beta \pm t\zeta \in C$ for all $t > 0$ sufficiently small$]$ is equivalent to $[\zeta \in Z]$, i.e., $Z = W_C(\beta)$. Thus $\#F_0 - 1 = \dim Z \leq n$. □

Appendix

Let V be a vector space over $\mathbf{R}$.

Theorem 1.A.1. *The space V is infinite-dimensional iff there exists a nonempty, convex subset C of V with icr $C = \emptyset$.*

PROOF. The "if" part is Theorem 1.4.1. To show the "only if" part, let B be a basis for V, well-ordered by $\leq$, such that it has no greatest element. Such order exists by Zermelo's theorem. One can identify V with $\oplus_B \mathbf{R}$, the set of all functions from B to $\mathbf{R}$ with a finite support. Define

$$C := \{\sum_{b \in B} \lambda_b b \mid \lambda \in \oplus_B \mathbf{R}. \text{ The "last" nonzero component is positive}\}.$$

Clearly, C is nonempty, proper, and convex. It is also clear that aff $C = V$.

To show that icr $C = \emptyset$, choose any $\sum_{b \in B} \lambda_b b \in C$. Let $\mu \in \oplus_B \mathbf{R}$ be such that its last nonzero component is negative (hence $\mu \notin C$), and such that the last nonzero component of μ comes "after" the last nonzero component of λ. Then

$$]\sum_{b \in B} \lambda_b b, \sum_{b \in B} \mu_b b] \cap C = \emptyset.$$

Hence
$$\sum_{b \in B} \lambda_b b \notin \text{icr } C. \quad \square$$

Theorem 1.A.2. *The space V is infinite-dimensional iff there exists a convex, proper subset C of V with $\lin C = V$.*

PROOF. *If.* Assume that $V \simeq \mathbf{R}^n$. Let C be a convex subset of $\mathbf{R}^n$ with $\lin C = \mathbf{R}^n$. It suffices to show that $\mathbf{R}^n \subset C$. Since $\lin C = \mathbf{R}^n$, $\aff C = \mathbf{R}^n$, therefore $\cor C \neq \emptyset$ by Theorem 1.4.1. Choose any $\bar{x} \in \cor C$. Then for any $x \in \mathbf{R}^n$, $2x - \bar{x} \in \lina C$. So
$$\exists y \in C : 2x - \bar{x} \neq y, \quad [y, 2x - \bar{x}[\subset C.$$
For this direction $2x - \bar{x} - y$,
$$\exists t > 0 : [\bar{x}, t(2x - \bar{x} - y) + \bar{x}] \subset C.$$
Then
$$x = \tfrac{1}{2}[(1 - t)(2x - \bar{x}) + ty] + \tfrac{1}{2}[t(2x - \bar{x} - y) + \bar{x}] \in C.$$

Only if. Consider the same example C as that in Theorem 1.A.1. Choose any $\sum_{b \in B} \lambda_b b \in V$. Take any $\sum_{b \in B} \mu_b b$ such that its last nonzero coordinate is positive and it comes "after" the last nonzero coordinate of $\sum \lambda_b b$. Then
$$]\sum_{b \in B} \lambda_b b, \sum_{b \in B} \mu_b b] \subset C. \quad \square$$

For further discussion see Klee (1951).

EXERCISES

1. Construct an example of a nonempty, closed subset S of $\mathbf{R}^n$ such that $\co S$ is an open, proper subset of $\mathbf{R}^n$.
2. Let $\{M_i\}_{i \in I}$ be an indexed family of affine subspaces of $\mathbf{R}^n$. Prove that $\bigcap_{i \in I} M_i$ is an affine subspace.

3. A function $T: \mathbf{R}^n \to \mathbf{R}^m$ is called *affine* if the map $x \mapsto T(x) - T(0)$ is linear. Prove that if T is affine, the image $T(C)$ of a convex set C in $\mathbf{R}^n$ is convex and the inverse image $T^{-1}(C')$ of a convex set C' in $\mathbf{R}^m$ is convex.

4. Let C be a convex set in $\mathbf{R}^n$. A function $f: C \to \mathbf{R}$ is called *convex on C* if the subset of $\mathbf{R}^n \times \mathbf{R}$ defined by

$$\mathrm{epi}\, f := \{(x, r) \in C \times \mathbf{R} \mid f(x) \le r\}$$

is convex. The set epi f is called the *epigraph of f*. Show that f is convex on C iff

$$f(\alpha x + (1-\alpha)y) \le \alpha f(x) + (1-\alpha)f(y)$$

for all $x, y \in C$ and all $\alpha: 0 \le \alpha \le 1$.

5. Let C be a convex subset of $\mathbf{R}^n$ with $\mathbf{0} \in \mathrm{cor}\, C$. The *Minkowski function of C* is a function $\rho_C: \mathbf{R}^n \to \mathbf{R}$, defined by

$$\rho_C(x) := \inf\{t > 0 \mid x \in tC\}.$$

Show that ρ_C is *sublinear* on $\mathbf{R}^n$ (i.e., $\forall x, y \in \mathbf{R}^n : \forall r \in \mathbf{R}_+ : \rho_C(rx) = r\rho_C(x)$, $\rho_C(x+y) \le \rho_C(x) + \rho_C(y)$); in particular, show that it is convex on $\mathbf{R}^n$.

6. Prove the *Hahn–Banach theorem*: Let W be a subspace of $\mathbf{R}^n$, let $f: \mathbf{R}^n \to \mathbf{R}$ be a convex function, and let $h: W \to \mathbf{R}$ be a linear function. If $h \le f|_W$, then there exists a linear extension h' to $\mathbf{R}^n$ of h such that $h' \le f$.
(*Hint*: Show that there exist $\bar{h} \in \mathbf{R}^n$, $\bar{k} \in \mathbf{R}$, and $t \in \mathbf{R}$ such that

$$(\bar{h}, \bar{k}) \ne (\mathbf{0}, 0),$$

$$\bar{h} \cdot x + \bar{k} \cdot r \ge t \quad \forall (x, r) \in \widehat{\mathrm{epi}\, f} \quad (\text{hence } \forall (x, r) \in \mathrm{epi}\, f),$$

$$\bar{h} \cdot x + \bar{k} \cdot r \le t \quad \forall (x, r) \in \mathrm{gr}\, h,$$

where gr h denotes

$$[\text{the } graph \text{ of } h] := \{(x, r) \in W \times \mathbf{R} \mid h(x) = r\}.$$

Explain that one may choose $t = 0$ so that

$$\mathrm{gr}\, h \subset H := \{(x, r) \in \mathbf{R}^n \times \mathbf{R} \mid \bar{h} \cdot x + \bar{k} \cdot r = 0\}$$

Fix any $r_0 \in \mathbf{R}$ such that $r_0 > f(0)$. Show that $(\mathbf{0}, r_0) \in$ cor epi f, and deduce from this that $\bar{k}(=\bar{h}\cdot\mathbf{0} + \bar{k}\cdot 1) > 0$. Define $h': \mathbf{R}^n \to \mathbf{R}$ by $h'(x) := -\bar{h}(x)/\bar{k}$ (so that $(h', -1)$ is a normal vector of H), and verify that h' satisfies the required properties.)

7. Let there be given a function $f: \mathbf{R}^n \to \mathbf{R}$ and a point $\bar{x} \in \mathbf{R}^n$. A linear map $h: \mathbf{R}^n \to \mathbf{R}$ is called a *subgradient of f at $\bar{x}$* if

$$\forall x \in \mathbf{R}^n : f(x) \geq f(\bar{x}) + h(x - \bar{x}).$$

Prove the *Subdifferentiability theorem*: Let $f: \mathbf{R}^n \to \mathbf{R}$ be a convex function, and let $\bar{x}$ be a point in $\mathbf{R}^n$. Then f has a subgradient at $\bar{x}$. (*Hint*: Consider a function $x \mapsto f(x + \bar{x}) - f(\bar{x})$ and a subspace $\{\mathbf{0}\}$. Apply the Hahn–Banach theorem.)

8. Use the subdifferentiability theorem to prove the support theorem. (*Hint*: Let C be a convex subset of $\mathbf{R}^n$, and let v be a point in $\mathbf{R}^n\backslash$icr C. Without loss of generality one may assume $\mathbf{0} \in$ cor C. Then the Minkowski function ρ_C of C is a convex function. Study a subgradient of ρ_C at v.)

REMARK. By the proof of the separation theorem (Theorem 1.5.4) and Exercises 6–8, the following cycle has been established:

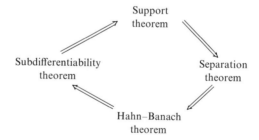

2

Introduction to Continuity of a Correspondence

Let X, Y be sets, and let $\mathscr{P}(Y)$ be the power set of Y. A correspondence from X to Y is a function from X to $\mathscr{P}(Y)$, i.e., a rule F that assigns to each point x of X a subset $F(x)$ of Y. Sometimes a correspondence is called a multivalued function. Given metrics on X and Y (or more generally, topologies on X and Y), one can define the continuity of a correspondence F by endowing $\mathscr{P}(Y)$ with a "natural" topology. Two of such topologies on $\mathscr{P}(Y)$ are introduced: the Vietoris topologies. The associated continuities of a correspondence F, called upper semicontinuity and lower semicontinuity, are studied and characterized (Lemmas 2.1.1 and 2.1.2). When Y is compact and $F(x)$ is closed in Y for each $x \in X$, upper semicontinuity of F is characterized as its closedness (Theorems 2.2.1 and 2.2.3). These concepts are applied in Section 2.3 to establish the fundamental theorem of the sensitivity analysis of optimization problems: the maximum theorem (Theorem 2.3.1). For a metric space Y one can introduce the Hausdorff distance on the family $\mathscr{P}'(Y) := \mathscr{P}(Y) \setminus \{\varnothing\}$. Two further continuities of F are defined, based on the Hausdorff distance, and their relationship with upper and lower semicontinuities is presented (Exercise 2). Finally, another continuity concept for F is introduced based on the topological limes superior (see Exercise 3). For a general reference relevant to the topological study of correspondences, see Berg (1959).

2.1. Upper and Lower Semicontinuities

Let X, Y be sets, fixed throughout this chapter, and denote by $\mathscr{P}(Y)$ the power set of Y. Given a subset S of $X \times Y$, one can define a function F from X to $\mathscr{P}(Y)$ by

$$F(x) := \{y \in Y \mid (x, y) \in S\}.$$

When there is no ambiguity, one simply writes $F: X \to Y$, and calls this a *correspondence* from X to Y. Sometimes a correspondence is called a *multivalued function*. The set S is called the *graph* of the correspondence F and is denoted by gr F. When F is regarded as a function to $\mathscr{P}(Y)$, the inverse of a member in $\mathscr{P}(Y)$ is defined as usual; namely

$$F^{-1}(B) := \{x \in X \mid F(x) = B\} \quad \text{for each} \quad B \in \mathscr{P}(Y).$$

The *inverse* F^{-1} is actually not handy, so different types of "inverses" are defined: For every $B \in \mathscr{P}(Y)$,

$$F^+(B) := \{x \in X \mid F(x) \subset B\},$$
$$F^-(B) := \{x \in X \mid F(x) \cap B \neq \varnothing\}.$$

The correspondences F^+, F^- are called the *upper* and *lower inverses* of F, respectively. The two inverses are related by $F^+(B) = X \backslash F^-(Y \backslash B)$. Clearly, for a non-empty-valued correspondence F

$$F^{-1}(B) \subset F^+(B) \subset F^-(B).$$

The following lemma is straightforward:

Lemma 2.1.1. *Let X, Y be metric spaces, and let F be a correspondence from X to Y. Then the following three conditions are equivalent:*

  (i) *For each open subset G of Y, its upper inverse $F^+(G)$ is open in X.*
  (ii) *For each closed subset C of Y, its lower inverse $F^-(C)$ is closed in X.*
  (iii) *For every sequence $\{x^k\}_k$ in X, converging to an arbitrarily given point $\bar{x}$, and for every neighborhood G of $F(\bar{x})$ in Y, there exists k_0 such that $F(x^k) \subset G$ for all $k \geq k_0$.*

2.1. Upper and Lower Semicontinuities

If one of the three conditions in the above lemma is satisfied the correspondence F is called *upper semicontinuous* (abbreviated henceforth as u.s.c.) *in* X. The correspondence F is called u.s.c. at $\bar{x}$ if for each neighborhood G of $F(\bar{x})$ in Y there exists a neighborhood U of $\bar{x}$ in X such that $F(x) \subset G$ for all $x \in U$.

Lemma 2.1.2. *Let X, Y be metric spaces, and let F be a correspondence from X to Y. Then the following three conditions are equivalent:*

(i) *For each open subset G of Y, its lower inverse $F^-(G)$ is open in X.*
(ii) *For each closed subset C of Y, its upper inverse $F^+(C)$ is closed in X.*
(iii) *For every sequence $\{x^k\}_k$ in X converging to an arbitrarily given $\bar{x}$, and for every open subset G of Y for which $F(\bar{x}) \cap G \neq \emptyset$ there exists k_0 such that $F(x^k) \cap G \neq \emptyset$ for all $k \geq k_0$.*

If one of the three conditions in this lemma is satisfied, the correspondence F is called *lower semicontinuous* (abbreviated henceforth as l.s.c.) *in* X. The correspondence F is called l.s.c. at $\bar{x}$ if for each open set G in Y for which $F(\bar{x}) \cap G \neq \emptyset$ there exists a neighborhood U of $\bar{x}$ in X such that $F(x) \cap G \neq \emptyset$ for all $x \in U$; this is the case if for every sequence $\{x^k\}_k$ in X converging to $\bar{x}$ and for every point y in $F(\bar{x})$, there exists a sequence $\{y^k\}_k$ in Y converging to y such that $y^k \in F(x^k)$ for all k sufficiently large (see Figure 2.1.1).

Given three sets X, Y, Z and two correspondences $F_1: X \to Y$ and $F_2: Y \to Z$, the *composite correspondence* $F_2 \circ F_1: X \to Z$ is defined by: $F_2 \circ F_1(x) := \bigcup_{y \in F_1(x)} F_2(y)$. The following corollary is straightforward by the identity $(F_2 \circ F_1)^- = F_1^- \circ F_2^-$:

Corollary 2.1.3. *Let X, Y, Z be metric spaces, and let $F_1: X \to Y$, $F_2: Y \to Z$ be correspondences. If F_1 and F_2 are both u.s.c. (both l.s.c., resp.), then $F_2 \circ F_1$ is u.s.c. (l.s.c., resp.).*

When a correspondence $F: X \to Y$ is single-valued, upper semicontinuity of F, lower semicontinuity of F, and continuity of F (as a function to Y) are all equivalent.

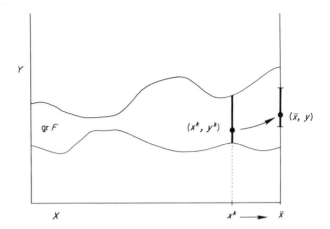

Figure 2.1.1 Lower semicontinuity at $\bar{x}$.

The concepts of upper and lower semicontinuities can also be defined for a correspondence F from a topological space X to a topological space Y. Lemmas 2.1.1 and 2.1.2 hold true in this general setup provided that the term "sequence" be replaced by the term "net" in condition (iii) of each lemma. Indeed, upper semicontinuity of F is equivalent to continuity of the *function* $F\colon X \to (\mathscr{P}(Y), \mathscr{U})$, where $\mathscr{U}$ is the topology on $\mathscr{P}(Y)$ generated by a subbase $\{\mathscr{P}(G) \mid G \text{ is open in } Y\}$. Notice the identity

$$\mathscr{P}(Y) \backslash \mathscr{P}(Y \backslash B) = \{C \in \mathscr{P}(Y) \mid C \cap B \neq \varnothing\}.$$

Lower semicontinuity of F is equivalent to continuity of the function $F\colon X \to (\mathscr{P}(Y), \mathscr{L})$ where $\mathscr{L}$ is the topology on $\mathscr{P}(Y)$ generated by a subbase $\{\mathscr{P}(Y) \backslash \mathscr{P}(Y \backslash G) \mid G \text{ is open in } Y\}$. The families $\mathscr{U}, \mathscr{L}$ are called the *Vietoris topologies*.

2.2. Closedness

Let X, Y be metric spaces. A correspondence $F\colon X \to Y$ is called *closed* if its graph gr F is closed in the product space $X \times Y$. Frequently

2.2. Closedness

it is easier to establish the closedness of F than to establish the upper semicontinuity of F.

Theorem 2.2.1. *Let X, Y be metric spaces, and let F be a correspondence from X to Y. If F is u.s.c. and closed-valued in X, then it is closed.*

PROOF. Choose any $(\bar{x}, \bar{y}) \notin \text{gr } F$. Then $\bar{y}$ is outside the closed set $F(\bar{x})$. There exist open sets V_1 and V_2 in Y such that $\bar{y} \in V_1$, $F(\bar{x}) \subset V_2$, and $V_1 \cap V_2 = \emptyset$. (Recall that a metric space is regular.) Since F is u.s.c., $F^+(V_2)$ is a neighborhood of $\bar{x}$. Then $F^+(V_2) \times V_1$ is a neighborhood of $(\bar{x}, \bar{y})$ disjoint from gr F. □

The converse of this theorem is false as the following example shows: Let

$$X = \mathbf{R}, \qquad Y = \mathbf{R}^2$$

and

$$F(r) = \{(x_1, x_2) \in \mathbf{R}^2 \mid x_2 \geq 0, (x_1 - r) \cdot x_2 \geq 1\}.$$

The correspondence F is closed, but is *not* u.s.c. at any point of $\mathbf{R}$. When the range Y is compact, however, the situation becomes pleasant. To see this one first has to establish

Lemma 2.2.2. *Let X, Y be metric spaces, and let F_i be correspondences from X to Y, $i = 1, 2$. Define a correspondence $F\colon X \to Y$ by $F(x) := F_1(x) \cap F_2(x)$. If F_1 is closed and if F_2 is u.s.c. and compact-valued in X, then F is u.s.c. in X.*

PROOF. Choose any $\bar{x} \in X$ and any neighborhood G of $F(\bar{x})$ in Y. If $F_2(\bar{x}) \subset G$, there is nothing to prove; so suppose $\exists y \in F_2(\bar{x}) \backslash G$. Note that $y \notin F_1(\bar{x})$; the point $(\bar{x}, y)$ is outside the closed set gr F_1. So there exist a neighborhood $N_y(\bar{x})$ of $\bar{x}$ in X and a neighborhood $V(y)$ of y in Y such that

$$[N_y(\bar{x}) \times V(y)] \cap \text{gr } F_1 = \emptyset,$$

i.e., $F_1(x) \cap V(y) = \emptyset$, for all $x \in N_y(\bar{x})$. Since the set $F_2(\bar{x}) \backslash G$ is compact, it has finitely many points $y^1, \ldots, y^n$ such that

$$V := V(y^1) \cup \cdots \cup V(y^n) \supset F_2(\bar{x}) \backslash G.$$

The set

$$N'(\bar{x}) := N_{y^1}(\bar{x}) \cap \cdots \cap N_{y^n}(\bar{x})$$

is a neighborhood of $\bar{x}$, and $F_1(x) \cap V = \emptyset$ for all $x \in N'(\bar{x})$. Since $G \cup V$ is a neighborhood of $F_2(\bar{x})$, there exists a neighborhood $N''(\bar{x})$ of $\bar{x}$ such that $F_2(x) \subset G \cup V$ for all $x \in N''(\bar{x})$. Then for all $x \in N'(\bar{x}) \cap N''(\bar{x})$, $F_1(x) \cap V = \emptyset$ and $F_2(x) \subset G \cup V$ so $F(x) \subset G$. □

Theorem 2.2.3. *Let X be a metric space, let Y be a compact metric space, and let F be a correspondence from X to Y. If F is closed, then it is u.s.c. and closed-valued in X.*

PROOF. Closed-valuedness is straightforward. To show upper semicontinuity define a constant correspondence $F_2: X \to Y$ by $F_2(x) = Y$ for all $x \in X$. This is u.s.c. and compact-valued. Now $F(x) = F(x) \cap F_2(x)$, so Lemma 2.2.2 is applicable. □

Note that the proofs of the preceding two theorems have also established the following: For a metric space X, a compact metric space Y, a correspondence $F: X \to Y$, and a point $\bar{x} \in X$ the following three conditions are equivalent.

(i) $F(\bar{x})$ is closed in Y, and F is u.s.c. at $\bar{x}$.
(ii) If $y \in Y \backslash F(\bar{x})$, then there exist a neighborhood N of $\bar{x}$ in X and a neighborhood V of y in Y such that $(N \times V) \cap \text{gr } F = \emptyset$.
(iii) If $\{x^k\}_k$ is a sequence in X converging to $\bar{x}$, $\{y^k\}_k$ is a sequence in Y converging to $\bar{y}$, and $y^k \in F(x^k)$ for all k, then $\bar{y} \in F(\bar{x})$. See Figure 2.2.1.

Theorem 2.2.1 can be generalized to the situation in which X is a topological space and Y is a regular topological space; Theorem 2.2.3 can be generalized to the situation in which X is a topological space and Y is a compact Hausdorff space.

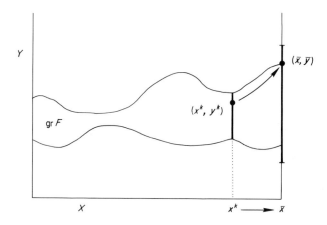

Figure 2.2.1 Closedness.

2.3. Maximum Theorem

Given a nonempty, compact set F_x and a continuous function $f_x: F_x \to \mathbf{R}$, the optimization problem

$$\text{maximize:} \quad f_x(y),$$
$$\text{subject to:} \quad y \in F_x, \qquad [\mathrm{P}_x]$$

has an optimal solution. When there is a set of optimization problems $[\mathrm{P}_x]$, parametrized by the x, a natural question to ask is how the optimal value and the optimal solution set for $[\mathrm{P}_x]$ change in response to a change in the parameter x.

Theorem 2.3.1 (Maximum Theorem). *Let X, Y be metric spaces, let $f: X \times Y \to \mathbf{R}$ be a continuous function, and let $F: X \to Y$ be a correspondence that is non-empty-valued, compact-valued, u.s.c., and l.s.c. in X. Then*

(i) *The function $\varphi: X \to \mathbf{R}$, $x \mapsto \max\{f(x, y) \mid y \in F(x)\}$ is continuous in X; and*

(ii) *The correspondence* $\Phi: X \to Y$, $x \mapsto \{y \in F(x) \mid f(x, y) = \varphi(x)\}$ *is u.s.c. in* X.

PROOF. To show the upper semicontinuity of Φ it suffices to show its closedness, since $\Phi(x) = \Phi(x) \cap F(x)$ and Lemma 2.2.2 applies. Suppose Φ is not closed; then there exists a sequence $\{x^k, y^k\}_k$ in gr Φ converging to $(\bar{x}, \bar{y}) \in (X \times Y) \backslash$ gr Φ. Clearly, $(\bar{x}, \bar{y}) \in$ gr F so $\exists y^* \in F(\bar{x}): f(\bar{x}, \bar{y}) < f(\bar{x}, y^*)$. By continuity of f there exist a neighborhood $U(\bar{x})$ of $\bar{x}$ in X, a neighborhood $V(y^*)$ of y^* in Y, and a positive number ϵ such that $f(\bar{x}, \bar{y}) + \epsilon < f(x, y)$ for all $(x, y) \in U(\bar{x}) \times V(y^*)$. Since F is l.s.c., $F(x^k) \cap V(y^*) \neq \emptyset$ for all k sufficiently large. Choose $y^{*k} \in F(x^k) \cap V(y^*)$. The sequence $\{x^k, y^{*k}\}_k$ will eventually be in gr $F \cap (U(\bar{x}) \times V(y^*))$—a contradiction to the definition of y^k since $f(x^k, y^k) \to f(\bar{x}, \bar{y})$.

Continuity of φ follows because φ is u.s.c. as the composite of the u.s.c. correspondence $x \mapsto \{x\} \times \Phi(x)$ and the u.s.c. (single-valued) correspondence $(x, y) \mapsto \{f(x, y)\}$ and because φ is single-valued. □

Note that the preceding proof has also established the following: Given metric spaces X, Y, a function $f: X \times Y \to \mathbf{R}$, a correspondence $F: X \to Y$, and a point $\bar{x} \in X$, if f is continuous at $(\bar{x}, y)$ for each $y \in F(\bar{x})$, $F(\bar{x})$ is nonempty and compact, and F is both u.s.c. and l.s.c. at $\bar{x}$, then the function φ (as defined in the preceding theorem) is continuous at $\bar{x}$ and the correspondence Φ is u.s.c. at $\bar{x}$.

The maximum theorem can be generalized to the situation in which X is a topological space and Y is a T_3 space.

EXERCISES

1. Consider a consumer characterized by his consumption set $X(\subset \mathbf{R}^l)$ and his utility function $u: X \to \mathbf{R}$. Let

$$\Delta^L := \{p \in \mathbf{R}^l \mid p \geq 0, \sum_{h=1}^{l} p_h = 1\}$$

be the price domain. The budget correspondence $\gamma: \Delta^L \times \mathbf{R}_+ \to X$ is defined by $\gamma(p, w) := \{x \in X \mid p \cdot x \leq w\}$, and the demand correspondence $\xi\colon \Delta^L \times \mathbf{R}_+ \to X$ is defined by

$$\xi(p, w) := \{x \in \gamma(p, w) \mid \forall y \in \gamma(p, w) : u(x) \geq u(y)\}.$$

Assume that X is compact and convex, and that u is continuous. Fix any $(\bar{p}, \bar{w}) \in \Delta^L \times \mathbf{R}_+$ for which $\min\{\bar{p} \cdot x \mid x \in X\} < \bar{w}$.

(i) Show that the budget correspondence γ is u.s.c. at $(\bar{p}, \bar{w})$. Choose any sequence $\{(p^k, w^k)\}_k$ in $\Delta^L \times \mathbf{R}_+$ that converges to $(\bar{p}, \bar{w})$, and choose any commodity bundle $\bar{x} \in \gamma(\bar{p}, \bar{w})$.

(ii) Show that $\exists k_0 : \forall k \geq k_0 : \gamma(p^k, w^k) \neq \emptyset$.

(iii) Assume that $\bar{p} \cdot \bar{x} < \bar{w}$. Construct for this case a sequence $\{x^k\}_k$ such that $x^k \in \gamma(p^k, w^k)$ for all k sufficiently large and that $x^k \to \bar{x}$.

Throughout (v)–(vii) assume that $\bar{p} \cdot \bar{x} = \bar{w}$.

(iv) Show that $\exists x' \in X : \bar{p} \cdot x' < \bar{w}$.

(v) Show that $\exists k_1 : \forall k \geq k_1 : p^k \cdot x' < w^k$ and $p^k \cdot x' < p^k \cdot \bar{x}$.

(vi) Show that $\forall k \geq k_1$: There exists uniquely a point $a^k \in \mathbf{R}^l$ such that $p^k \cdot a^k = w^k$ and $\{x', \bar{x}, a^k\}$ is on one line.

(vii) Define for all $k \geq k_1$

$$x^k = \begin{cases} a^k & \text{if } a^k \in [x', \bar{x}], \\ \bar{x} & \text{if } \bar{x} \in [x', a^k], \end{cases}$$

and show that $x^k \in \gamma(p^k, w^k)$ for all k sufficiently large and that $x^k \to \bar{x}$.

(viii) Conclude that γ is l.s.c. at $(\bar{p}, \bar{w})$.

(ix) Show that the demand correspondence is u.s.c. at $(\bar{p}, \bar{w})$.

2. Let $X, (Y, d)$ be metric spaces and $\mathscr{P}'(Y)$ the family of nonempty subsets of Y, $\mathscr{P}(Y)\setminus\{\emptyset\}$. Given two members B, C of $\mathscr{P}'(Y)$, define

$$\rho(B, C) := \sup_{y \in B} \inf_{z \in C} d(y, z),$$

$$D(B, C) := \max[\rho(B, C), \rho(C, B)].$$

The function $D \colon \mathscr{P}'(Y) \times \mathscr{P}'(Y) \to [0, \infty]$ satisfies all the properties of a pseudometric for $\mathscr{P}'(Y)$ except that it may be infinite-valued. (When (Y, d) is a metric space of finite diameter, then the restriction of D to $\{B \in \mathscr{P}'(Y) \mid B \text{ is closed in } (Y, d)\}$ is called the

Hausdorff metric.) But D defines the same topology as a pseudometric $\min[D, 1]$.

A correspondence $F: X \to Y$ is called *H-upper semicontinuous* (abbreviated henceforth as *H-u.s.c.*) *at* $\bar{x} \in X$ if for each sequence $\{x^k\}_k$ in X converging to $\bar{x}$ it follows that $\{\rho(F(x^k), F(\bar{x}))\}_k$ converges to 0. A correspondence $F: X \to Y$ is called *H-lower semicontinuous* (*H-l.s.c.*) *at* $\bar{x} \in X$, if for each sequence $\{x^k\}_k$ in X converging to $\bar{x}$ it follows that $\{\rho(F(\bar{x}), F(x^k))\}_k$ converges to 0. A correspondence $F: X \to Y$ is called *H-u.s.c.* (*H-l.s.c.*, resp.) *in* X if it is H-u.s.c. (H-l.s.c., resp.) at every point of X.

To avoid ambiguity upper semicontinuity and lower semicontinuity, as defined in Section 2.1, are called here *V-upper semicontinuity and V-lower semicontinuity*, respectively.

(i) Prove that if $F: X \to Y$ is a non-empty-valued correspondence and is V-u.s.c. at $\bar{x} \in X$, then F is H-u.s.c. at $\bar{x}$.

(ii) Prove that if $F: X \to Y$ is a non-empty-valued correspondence such that it is H-u.s.c. at $\bar{x} \in X$ and $F(\bar{x})$ is compact, then F is V-u.s.c. at $\bar{x}$.

(iii) Construct an example of a non-empty-valued correspondence, $F: X \to Y$, that is not V-u.s.c. at $\bar{x} \in X$ and yet is H-u.s.c. at $\bar{x} \in X$.

(iv) Prove that if $F: X \to Y$ is a non-empty-valued correspondence and is H-l.s.c. at $\bar{x} \in X$, then F is V-l.s.c. at $\bar{x}$.

(v) Prove that if $F: X \to Y$ is a non-empty-valued correspondence such that it is V-l.s.c. at $\bar{x} \in X$ and $F(\bar{x})$ is totally bounded, then F is H-l.s.c. at $\bar{x}$.

(vi) Construct an example of a non-empty-valued correspondence, $F: X \to Y$, that is not H-l.s.c. at $\bar{x} \in X$ and yet is V-l.s.c. at $\bar{x}$.

3. Let X and Y be metric spaces, and let $\mathscr{P}(Y)$ be the power set of Y. Let $\{T^k\}_k$ be a sequence in $\mathscr{P}(Y)$. The *topological limes superior* of $\{T^k\}_k$ is the set $\mathrm{ls}(T^k) := \{y \in Y \mid \text{For every neighborhood } V \text{ of } y \text{ in } Y \text{ there exist infinitely many } k \text{ with } V \cap T^k \neq \varnothing.\}$. The *topological limes inferior* of $\{T^k\}_k$ is the set $\mathrm{li}(T^k) := \{y \in Y \mid \text{For every neighborhood } V \text{ of } y \text{ in } Y \text{ there exists } k_V \text{ such that } V \cap T^k \neq \varnothing \text{ for all } k \geq k_V.\}$. Clearly both $\mathrm{ls}(T^k)$ and $\mathrm{li}(T^k)$ are closed in Y, and

li(T^k) ⊂ ls(T^k). A correspondence $F: X \to Y$ is called *H–K-upper semi-continuous* if for each sequence $\{x^k\}_k$ in X converging to $\bar{x}$ it follows that ls($F(x^k)$) ⊂ $F(\bar{x})$. (Here H–K stands for Hahn–Kuratowski.) Prove that the following three conditions are equivalent:

(i) The correspondence F is H–K-upper semicontinuous.
(ii) For each sequence $\{x^k\}_k$ in X converging to $\bar{x}$ it follows that li($F(x^k)$) ⊂ $F(\bar{x})$.
(iii) The correspondence F is closed.

REMARK. The equivalence in Exercise 3 is no longer true within a more general framework of topological spaces. Within the metric space framework, the following implications have been established for a non-empty-valued correspondence F:

F is H–K-u.s.c.
⇕
F is closed.
⇑
F is V-u.s.c. and closed-valued.
⇓
F is V-u.s.c.
⇓
F is H-u.s.c.

F is V-l.s.c.
⇑
F is H-l.s.c.

Moreover,

F is closed
⇓ if Y is compact.
F is V-u.s.c. and closed-valued.
⇑ if F is compact-valued.
F is H-u.s.c.

F is V-l.s.c.
⇓ if F is totally bounded-valued.
F is H-l.s.c.

3

Introduction to Fixed-Point Theorems in $\mathbf{R}^n$

Two fixed-point theorems are presented: Brouwer's fixed-point theorem (Theorem 3.2.1) and Kakutani's fixed-point theorem (Theorem 3.2.2). The proof of Theorem 3.2.1 in the text follows the classical technique: First, one establishes Sperner's lemma, then uses the lemma to establish the Knaster–Kuratowski–Mazurkiewicz theorem (henceforth called the K–K–M theorem), and then finally uses the K–K–M theorem to prove Theorem 3.2.1. The fixed-point theorem (Theorem 3.2.1) can be, in turn, used to prove the K–K–M theorem. The proof of Theorem 3.2.2 in the text follows the original technique by Kakutani. Also presented are Michael's theorem on a continuous selection (Exercise 2 (ii)) and Cellina's technique to prove Kakutani's fixed-point theorem via the Michael theorem (Exercise 3). In Section 3.3 Ky Fan's coincidence theorem is presented (Theorem 3.3.3). The theorem can be considered to be a synthesis of Kakutani's fixed-point theorem and the separation principle.

3.1. Sperner's Lemma, K–K–M Theorem

Let $(x^i)_{i \in F}$ be an affinely independent subset of $\mathbf{R}^n$, with $\#F = k + 1$, fixed throughout this section. Given $F' \subset F$, the simplex $\mathrm{co}(x^i)_{i \in F'}$ is called a ($\#F' - 1$)-*dimensional face* of the simplex $\text{co}(x^i)_{i \in F}$. A collection of simplexes $\mathscr{S}$ is called a *simplicial partition* of $\text{co}(x^i)_{i \in F}$ if (1) $\bigcup_{S \in \mathscr{S}} S = \text{co}(x^i)_{i \in F}$ and (2) $[S, S' \in \mathscr{S}]$ implies $[S \cap S' = \emptyset$, or $S \cap S'$ is a face both of S and of $S']$. A *vertex* of $\mathscr{S}$ is a vertex of some member of $\mathscr{S}$.

Lemma 3.1.1 (Sperner). *Let $(x^i)_{i \in F}$ be an affinely independent subset of $\mathbf{R}^n$, with $\#F = k + 1$, $\mathscr{S}$ a simplicial partition of $\text{co}(x^i)_{i \in F}$, and let f be a function from {the vertices of $\mathscr{S}$} into $(x^i)_{i \in F}$. Assume that $[x$ is a vertex of $\mathscr{S}$, $F' \subset F$, $x \in \text{co}(x^i)_{i \in F'}]$ implies $[f(x) \in (x^i)_{i \in F'}]$. Then there exists $S \in \mathscr{S}$ for which f(the vertices of S) $= (x^i)_{i \in F}$.*

The assertion of Sperner's lemma is illustrated in Figure 3.1.1. The shaded subsimplexes are those S in $\mathscr{S}$ for which f(the vertices of S) = $\{a, b, c\}$; they are called *completely labeled*. One may even conjecture that the number of completely labeled subsimplexes is odd; it is indeed the case as the following proof of Sperner's lemma verifies.

Before presenting a proof of Sperner's lemma with its full generality, an alternative proof for the case $n = 2$ is given. The alternative proof follows the path-following technique of Lemke and Howson. Add a new vertex with the label "a," and join this new vertex with each vertex of $\mathscr{S}$

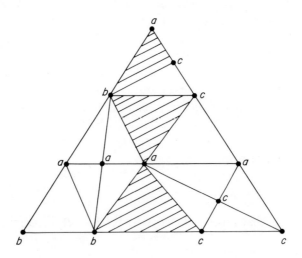

Figure 3.1.1

3.1. Sperner's Lemma, K–K–M Theorem

that lies in the face $[a, b]$ of the original simplex co$\{a, b, c\}$ (see Figure 3.1.2). One moves from a subsimplex to an adjacent subsimplex by crossing over the edge labeled $[a, b]$; one cannot cross over the other edges. The starting position is marked in the figure as "Start." For each subsimplex during the move either there are two edges labeled $[a, b]$, or there is only one edge labeled $[a, b]$; in the latter case the subsimplex is completely labeled. One can return to none of the subsimplexes already passed through, for if one could, there would be a subsimplex all three edges of which were labeled $[a, b]$; but that is impossible. There is no way to get outside the picture, since the starting edge is the only one labeled $[a, b]$ that is on the boundary of the picture. There are only finitely many subsimplexes. Thus the move stops after finitely many stages; the stopping position is the required completely labeled subsimplex. To show that the number of the completely labeled subsimplexes is odd, just note that any completely labeled subsimplex other than the one arrived at in Figure 3.1.2 has a "partner completely labeled subsimplex"; see Figure 3.1.3.

PROOF OF LEMMA 3.1.1. It will be shown by induction on k that the number of simplexes having the property asserted in the lemma is odd. This is trivially true for $k = 0$. Assume that this is true also for $k - 1$,

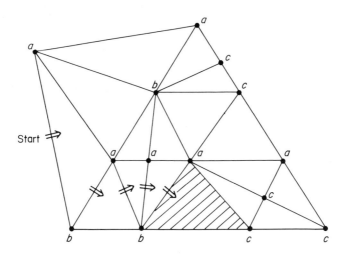

Figure 3.1.2

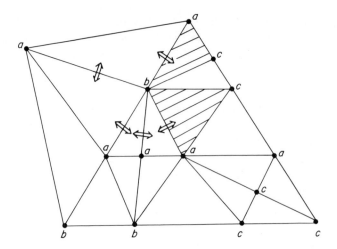

Figure 3.1.3

and consider $\mathscr{S}$ and f that satisfy the assumptions of the lemma. Choose any $F' \subset F$, with $\#F' = k$. For each $S \in \mathscr{S}$, let r_S be the number of $(k-1)$-dimensional faces each of whose vertices is mapped under f onto $(x^i)_{i \in F'}$. Then S has the property asserted in the lemma iff $r_S = 1$; otherwise r_S is equal to 0 or 2. So [the number of simplexes having the property asserted in the lemma] $\equiv \sum_{S \in \mathscr{S}} r_S \pmod{2}$. It suffices to show that $\sum_{S \in \mathscr{S}} r_S$ is odd.

Let T be any $(k-1)$-dimensional face of a member of $\mathscr{S}$ whose vertices are mapped onto $(x^i)_{i \in F'}$. If T is not contained in any $(k-1)$-dimensional face of $\mathrm{co}(x^i)_{i \in F}$, then it is the intersection of two adjacent members of $\mathscr{S}$ of dimension k, so it is counted twice in the sum $\sum_{S \in \mathscr{S}} r_S$. Therefore $\sum_{S \in \mathscr{S}} r_S$ is odd iff the number of such T contained in any $(k-1)$-dimensional face of $\mathrm{co}(x^i)_{i \in F}$ is odd. Now if T is contained in any $(k-1)$-dimensional face of $\mathrm{co}(x^i)_{i \in F}$, then that $(k-1)$-dimensional face should be $\mathrm{co}(x^i)_{i \in F'}$ because of the assumption on f. The inductive hypothesis is now applicable. □

An immediate consequence of Sperner's lemma is the following theorem (henceforth called the K–K–M theorem), which is actually equivalent to a fixed-point theorem as will be seen in the next section.

Theorem 3.1.2 (Knaster–Kuratowski–Mazurkiewicz). *Let $(x^i)_{i \in F}$ be an affinely independent subset of $\mathbf{R}^n$, and let $\{C^i\}_{i \in F}$ be a family of closed subsets of the simplex $\operatorname{co}(x^i)_{i \in F}$ such that for each $F' \subset F$ it follows that $\operatorname{co}(x^i)_{i \in F'} \subset \bigcup_{i \in F'} C^i$. Then $\bigcap_{i \in F} C^i \neq \emptyset$.*

PROOF. Let $\{\mathscr{S}^q\}_q$ be a sequence of simplicial partitions of $\operatorname{co}(x^i)_{i \in F}$ such that $d^q := \max\{\text{the diameter of } S \mid S \in \mathscr{S}^q\} \to 0$ as $q \to \infty$. For each q define a function f^q from $\{$the vertices of $\mathscr{S}^q\}$ into $(x^i)_{i \in F}$ as follows. For each vertex x of $\mathscr{S}^q$ there exists a unique subset F' of F such that $x \in \operatorname{icr} \operatorname{co}(x^i)_{i \in F'}$. Since $\operatorname{co}(x^i)_{i \in F'} \subset \bigcup_{i \in F'} C^i$, there exists $i \in F'$ such that $x \in C^i$. Choose any such i, and set $f^q(x) := x^i$. Then $\mathscr{S}^q, f^q$ satisfy the assumptions of Sperner's lemma; there exists $S^q \in \mathscr{S}^q$ such that

$$f^q(\text{the vertices of } S^q) = (x^i)_{i \in F}.$$

Let $(x^{i,q})_{i \in F}$ be the vertices of S^q; w.l.o.g. $f^q(x^{i,q}) = x^i$, hence $x^{i,q} \in C^i$ for each $i \in F$. Passing through a subsequence if necessary, one can assume w.l.o.g. that the sequence $\{x^{i,q}\}_q$ converges to some point, say $\bar{x}^i$, for each i. Since C^i is closed, $\bar{x}^i \in C^i$. Since $d^q \to 0$, all the $\bar{x}^i$ are identical to a point, say x^*. Thus $x^* \in \bigcap_{i \in F} C^i$. □

3.2. Fixed-Point Theorems

Theorem 3.2.1 (Brouwer's Fixed-Point Theorem). *Let C be a nonempty, convex, compact subset of $\mathbf{R}^n$, and let $f: C \to C$ be a continuous function from C to itself. Then f has a fixed-point, i.e., there exists $x^* \in C$ such that $x^* = f(x^*)$.*

PROOF. *Step 1.* Assume C is a simplex, i.e., the convex hull of an affinely independent set $(x^i)_{i \in F}$. Each point $x \in C$ is uniquely expressed as a convex combination of the vertices $x = \sum_{i \in F} \alpha_i(x) x^i$, $\alpha_i(x) \geq 0$, $\sum_{i \in F} \alpha_i(x) = 1$. For each $i \in F$ define $C^i := \{x \in C \mid \alpha_i(f(x)) \leq \alpha_i(x)\}$. Each C^i is closed by continuity of f. To show that the family $\{C^i\}_{i \in F}$ satisfies the assumption in the K–K–M theorem, choose any subset F'

of F and any point $x \in \text{co}(x^i)_{i \in F'}$. Then $\sum_{i \in F'} \alpha_i(x) = 1 \geq \sum_{i \in F'} \alpha_i(f(x))$ so $\exists i \in F': \alpha_i(f(x)) \leq \alpha_i(x)$. Consequently $\text{co}(x^i)_{i \in F'} \subset \bigcup_{i \in F'} C^i$. Thus by the K–K–M theorem there exists $x^* \in C$ such that $\alpha_i(f(x^*)) \leq \alpha_i(x^*)$ for all $i \in F$. Since $\sum_{i \in F} \alpha_i(f(x^*)) = \sum_{i \in F} \alpha_i(x^*) = 1$, it follows that $\alpha_i(f(x^*)) = \alpha_i(x^*)$ for all $i \in F$; i.e., $f(x^*) = x^*$.

Step 2. Let C be a nonempty, convex, compact subset of $\mathbf{R}^n$, with aff C having dimension k. There exists an affinely independent set $(x^i)_{i \in F}$, with $\#F = k + 1$, such that $C \subset \text{co}(x^i)_{i \in F}$. Choose any $\bar{x} \in \text{icr } C$, and extend f to $\text{co}(x^i)_{i \in F}$ by

$$\alpha(x) := \max\{\alpha \in \mathbf{R} \mid 0 \leq \alpha \leq 1, (1 - \alpha)\bar{x} + \alpha x \in C\},$$

$$f'(x) := f((1 - \alpha(x))\bar{x} + \alpha(x)x)$$

for every $x \in \text{co}(x^i)_{i \in F}$. The function α is continuous by convexity of C so the extension f' is continuous. By Step 1, the extension f' has a fixed-point in $\text{co}(x^i)_{i \in F}$. The fixed-point is actually in C because $f'(\text{co}(x^i)_{i \in F}) \subset C$; hence it is a fixed-point of f. □

Theorem 3.2.2 (Kakutani's Fixed-Point Theorem). *Let C be a nonempty, convex, compact subset of $\mathbf{R}^n$, and let $F: C \to C$ be an u.s.c., non-empty-valued, closed-valued, and convex-valued correspondence. Then F has a fixed-point, i.e., there exists $x^* \in C$ such that $x^* \in F(x^*)$.*

PROOF. *Step 1.* Assume that C is a simplex, i.e., the convex hull of an affinely independent set $(x^i)_{i \in G}$. Let $\{\mathcal{S}^q\}_q$ be a sequence of simplicial partitions of C such that $d^q := \max\{\text{the diameter of } S \mid S \in \mathcal{S}^q\} \to 0$ as $q \to \infty$. For each q a continuous function $f^q: C \to C$ is now constructed. If x is a vertex of $\mathcal{S}^q$, choose any point y in $F(x)$ and set $f^q(x) := y$. Then extend f^q piecewise linearly to C; that is, if $x \in S \in \mathcal{S}^q$ and $(x^{i,S})_{i \in G}$ are the vertices of S so that $x = \sum_{i \in G} \alpha_i x^{i,S}$, with $\alpha_i \geq 0$ and $\sum_{i \in G} \alpha_i = 1$, then $f^q(x) := \sum_{i \in G} \alpha_i f^q(x^{i,S})$. By Brouwer's fixed-point theorem, there exists $x^q \in C$ such that $x^q = f^q(x^q)$. Let S^q be a $(\#G - 1)$-dimensional simplex of $\mathcal{S}^q$ that contains x^q, and let $(x^{i,q})_{i \in G}$ be the vertices of S^q. Then

$$x^q = \sum_{i \in G} \alpha_i^q x^{i,q}, \qquad \alpha_i^q \geq 0, \qquad \sum_{i \in G} \alpha_i^q = 1.$$

Passing through a subsequence if necessary, one can assume w.l.o.g.

3.2. Fixed-Point Theorems

that for each $i \in G$ the sequence $\{\alpha_i^q\}_q$ converges to α_i^* and the sequence $\{f^q(x^{i,q})\}_q$ converges to y^{i*}. Clearly, $\alpha_i^* \geq 0$ and $\sum_{i \in G} \alpha_i^* = 1$. Since $d^q \to 0$, one may also assume w.l.o.g., that the sequences $\{x^q\}_q, \{x^{i,q}\}_q$, $i \in G$, all have the same limit $x^* \in C$. Now for each i, $(x^{i,q}, f^q(x^{i,q})) \in \mathrm{gr}\, F$ for all q, so $y^{i*} \in F(x^*)$. Also, by the definition of f^q,

$$x^q = f^q(x^q) = \sum_{i \in G} \alpha_i^q f^q(x^{i,q}),$$

so

$$x^* = \sum_{i \in G} \alpha_i^* y^{i*} \in \mathrm{co}\, F(x^*) = F(x^*).$$

Step 2. Proof for the general case proceeds exactly as in Step 2 of the proof of Brouwer's fixed-point theorem: One considers a simplex $\mathrm{co}(x^i)_{i \in G}$ that contains C, defines a continuous function $\alpha \colon \mathrm{co}(x^i)_{i \in G} \to [0, 1]$, and obtain an u.s.c. extension F' of F to $\mathrm{co}(x^i)_{i \in G}$ by setting $F'(x) := F((1 - \alpha(x))\bar{x} + \alpha(x)x)$. □

A forerunner to theorem 3.2.2 is found in von Neumann (1937). Clearly, the Kakutani theorem can be specialized to the Brouwer theorem when F is single-valued. The present proofs of these two theorems, therefore, have identified the logic: The K–K–M theorem $\Rightarrow$ Brouwer's fixed-point theorem $\Leftrightarrow$ Kakutani's fixed-point theorem. One can close this "cycle" by using Brouwer's fixed-point theorem to prove the K–K–M theorem. The following is one of such proofs and is due to Kim Border and Edward J. Green.

ALTERNATIVE PROOF OF THE K–K–M THEOREM. The same notation as that in the statement of the K–K–M theorem (Theorem 3.1.2) is used here. Suppose $\bigcap_{i \in F} C^i = \emptyset$, and set $D^i := \mathrm{co}(x^i)_{i \in F} \setminus C^i$. Then the family $\{D^i\}_{i \in F}$ is an open cover of the compact space $\mathrm{co}(x^i)_{i \in F}$. Let $\{g_i\}_{i \in F}$ be a partition of unity subordinate to $\{D^i\}_{i \in F}$; i.e., $g_i \colon \mathrm{co}(x^i)_{i \in F} \to \mathbf{R}$ is continuous, $g_i(x) \geq 0$ and $\sum_{i \in F} g_i(x) = 1$ for all $x \in \mathrm{co}(x^i)_{i \in F}$, and $g_i(x) = 0$ for all $x \notin D^i$. Define $g \colon \mathrm{co}(x^i)_{i \in F} \to \mathrm{co}(x^i)_{i \in F}$ by $g(x) := \sum_{i \in F} g_i(x) x^i$. By Brouwer's fixed-point theorem, there exists $x^* \in \mathrm{co}(x^i)_{i \in F}$ such that $x^* = g(x^*)$. Set $F' := \{i \in F \mid g_i(x^*) > 0\}$. The set F' is nonempty, and $x^* \notin C^i$ for all $i \in F'$ so $x^* \notin \bigcup_{i \in F'} C^i$. However, $x^* = \sum_{i \in F'} g_i(x^*) x^i \in \mathrm{co}(x^i)_{i \in F'}$. The nonemptiness of the set $\mathrm{co}(x^i)_{i \in F'} \setminus \bigcup_{i \in F'} C^i$ contradicts the assumption in the K–K–M theorem. □

Kakutani's fixed-point theorem (Brouwer's fixed-point theorem, resp.) can be generalized to the situation in which the Euclidean space $\mathbf{R}^n$ is replaced by a locally convex Hausdorff topological vector space. This generalization was established independently by Ky Fan (1952) and I. L. Glicksberg (1952) (by A. Tychonoff 1935s, resp.).

3.3. Fixed-Point Theorem and Separation Principle: Coincidence Theorem

In the late 1960s Ky Fan established a result that is much sharper than Kakutani's fixed-point theorem. A simpler version of this result is presented here; the version can be considered as a synthesis of Kakutani's fixed-point theorem and the separation principle.

Lemma 3.3.1 (Browder–Karamardian). *Let C be a nonempty, convex, compact subset of $\mathbf{R}^n$, and let $p: C \to \mathbf{R}^n$ be a continuous function. Then there exists $x^* \in C$ such that $p(x^*) \cdot x^* \leq p(x^*) \cdot x$ for all $x \in C$.*

PROOF. For each $x \in C$ define $\Phi(x) := \{ y \in C \mid \forall y' \in C : p(x) \cdot y \leq p(x) \cdot y' \}$. Clearly, $\Phi(x)$ is nonempty and convex. By the maximum theorem, the correspondence $x \mapsto \Phi(x)$ is u.s.c. in C. By Kakutani's fixed-point theorem, there exists $x^* \in C$ such that $x^* \in \Phi(x^*)$, i.e., such that $p(x^*) \cdot x^* \leq p(x^*) \cdot y'$ for all $y' \in C$. □

The point x^* whose existence is asserted in Lemma 3.3.1 is customarily called a *stationary point* of C. The existence of and the algorithm for stationary points have been studied as a complementarity problem in operations research literature.

Theorem 3.3.2 (Ky Fan). *Let C be a nonempty, convex, compact subset of $\mathbf{R}^n$, and let F, G be two u.s.c., non-empty-valued correspondences from C to $\mathbf{R}^n$ such that*

3.3. Coincidence Theorem

(∗) *For any $x \in C$ and any $p \in \mathbf{R}^n$ for which $p \cdot x = \min\{p \cdot y \mid y \in C\}$, there exist $u \in F(x)$ and $v \in G(x)$ such that $p \cdot u \geq p \cdot v$.*

Then there exists $x^* \in C$ for which $F(x^*)$ and $G(x^*)$ cannot be strictly separated by an hyperplane, i.e., for any $p \in \mathbf{R}^n$ and any $t \in \mathbf{R}$ it is not true that $p \cdot u < t$ for all $u \in F(x^*)$ and $p \cdot v > t$ for all $v \in G(x^*)$.

PROOF. For each $p \in \mathbf{R}^n$ define

$$P(p) := \{x \in C \mid \forall u \in F(x) : \forall v \in G(x) : p \cdot u < p \cdot v\},$$

and denote by $\overset{\circ}{P}(p)$ the interior of $P(p)$ in C. Suppose the assertion of the theorem is false; for any $x \in C$ there exist $p_x \in \mathbf{R}^n$ and $t_x \in \mathbf{R}$ such that $F(x) \subset \{u \in \mathbf{R}^n \mid p_x \cdot u < t_x\}$ and $G(x) \subset \{v \in \mathbf{R}^n \mid p_x \cdot v > t_x\}$. By upper semicontinuity of F and G, there exists a neighborhood $N(x)$ of x in C such that for all $x' \in N(x)$ it follows that $F(x') \subset \{u \in \mathbf{R}^n \mid p_x \cdot u < t_x\}$ and $G(x') \subset \{v \in \mathbf{R}^n \mid p_x \cdot v > t_x\}$; hence $x \in \overset{\circ}{P}(p_x)$. Thus $\{\overset{\circ}{P}(p_x) \mid x \in C\}$ is an open cover of C, so it has a finite subcover $\{\overset{\circ}{P}(p_x) \mid x \in I\}$, with $\#I < \infty$. Let $\{g_x\}_{x \in I}$ be a partition of unity subordinate to $\{\overset{\circ}{P}(p_x) \mid x \in I\}$, and define a continuous function $p: C \to \mathbf{R}^n$ by $p(y) := \sum_{x \in I} g_x(y) p_x$. Then for all $y \in C$, $u \in F(y)$, $v \in G(y)$, it follows that $p(y) \cdot u < p(y) \cdot v$. By Lemma 3.3.1, however, there exists $y^* \in C$ such that $p(y^*) \cdot y^* \leq p(y^*) \cdot y$ for all $y \in C$, so in view of the assumption (∗), there exist $u^* \in F(y^*)$ and $v^* \in G(y^*)$ such that $p(y^*) \cdot u^* \geq p(y^*) \cdot v^*$—a contradiction. □

Kakutani's fixed-point theorem was reformulated in Theorem 3.3.2 in such a way that one can easily introduce the separation principle for a further result.

Theorem 3.3.3 (Ky Fan's Coincidence Theorem). *Let C, F, G be defined as in Theorem 3.3.2. In particular, assume nonemptiness, convexity, and compactness of C; assume upper semicontinuity, non-empty-valuedness of F and G; and assume condition (∗). Further assume that for each $x \in C$ both $F(x)$ and $G(x)$ are closed, convex subsets of $\mathbf{R}^n$ and that at least one of them is compact. Then there exists $x^* \in C$ for which $F(x^*) \cap G(x^*) \neq \varnothing$.*

PROOF. If for each $x \in C$, $F(x) \cap G(x) = \varnothing$, then there exists a hyperplace that strictly separates $F(x)$ and $G(x)$. This contradicts the assertion of the preceding theorem. □

The coincidence theorem (Theorem 3.3.3) was proved by applying Kakutani's fixed-point theorem and the separation principle. The latter two results, in turn, can be proved by applying the coincidence theorem:

ALTERNATIVE PROOF OF KAKUTANI'S FIXED-POINT THEOREM. Let C be a nonempty, convex, compact subset of $\mathbf{R}^n$, and let $F: C \to C$ be an u.s.c., non-empty-valued, closed-valued, and convex-valued correspondence. Define a (single-valued) correspondence $G: C \to C$ by $G(x) = \{x\}$. It is straightforward to check that all of the assumptions in the coincidence theorem are satisfied, so there exists $x^* \in C$ such that $F(x^*) \cap \{x^*\} \neq \emptyset$. □

ALTERNATIVE PROOF OF THE SEPARATION PRINCIPLE. Let S be a nonempty, closed, convex subset of $\mathbf{R}^n$, and let v be a point in $\mathbf{R}^n \setminus S$. Let F, G be correspondences from $\{0\}$ to $\mathbf{R}^n$ defined by $F(0) := S$ and $G(0) := \{v\}$. Then $\{0\}, F, G$ satisfy all the assumptions of the coincidence theorem, except possibly (*), and the assertion of the coincidence theorem is false, i.e., $F(0) \cap G(0) = \emptyset$. Therefore the negation of (*) holds true; there exists $p \in \mathbf{R}^n$ such that for any $u \in S$ it follows that $p \cdot u < p \cdot v$. □

Ky Fan established the preceding two theorems in the infinite-dimensional context. Theorem 3.3.2 (Theorem 3.3.3, resp.) holds true even when $\mathbf{R}^n$ is replaced by a Hausdorff topological vector space E over $\mathbf{R}$ (by a Hausdorff, locally convex topological vector space E over $\mathbf{R}$, resp.); here, the vector p in the condition (*) is understood to be a member of the topological dual E^* of E. As can be seen from the proofs, upper semicontinuity of the correspondences F, G in both theorems may also be replaced by a weaker condition of their upper demicontinuity. A correspondence F from a topological space C to a topological vector space E is called *upper demicontinuous* (abbreviated henceforth as u.d.c.), if for each $x \in C$, each $p \in E^*$, and each $t \in \mathbf{R}$ for which $p \cdot y < t$ for all $y \in F(x)$, it follows that there exists a neighborhood $U(x)$ of x in C such that $p \cdot y < t$ for all $y \in \bigcup_{x' \in U(x)} F(x')$. This last

Exercises

generalization has proven substantially useful, as the following discussion suggests.

Let F' be a non-empty-valued, compact-valued correspondence from a topological space C to a topological vector space E. Define a correspondence $F: C \to E$ by $F(x) := \overline{\text{co}}\, F'(x)$, the closed convex hull of $F'(x)$, i.e., the smallest closed convex set that contains $F'(x)$. If E is a Euclidean space, then upper semicontinuity of F' implies upper semicontinuity of F (by Carathéodory's theorem). If E is infinite-dimensional, however, it is not known whether the assertion holds true. Now upper demicontinuity of F' (in particular, upper semicontinuity of F') implies upper demicontinuity of F, regardless of the dimension of E.

EXERCISES

1. A subspace S of a topological space X is called a *retract* of X if the identity map id: $S \to S$, $x \mapsto x$, has a continuous extension $f: X \to S$. Define

$$B^n := \{x \in \mathbf{R}^n \mid \|x\| \leq 1\},$$

$$S^{n-1} := \{x \in \mathbf{R}^n \mid \|x\| = 1\}.$$

Consider the following two statements:

(B) Any continuous function $f: B^n \to B^n$ has a fixed-point.
(R) S^{n-1} is not a retract of B^n.

Use (B) to prove (R); conversely, use (R) to prove (B).

2. Let X be a compact metric space, and let $F: X \to \mathbf{R}^m$ be a l.s.c., non-empty-valued, and convex-valued correspondence.
    (i) Prove that for every $\epsilon > 0$ there exists a continuous function $f: X \to \mathbf{R}^m$ such that $\forall x \in X: \rho(\{f(x)\}, F(x)) < \epsilon$, where ρ is defined as in Exercise 2 of Chapter 2. (*Hint*: For each $y \in \mathbf{R}^m$ show that $U(y) := \{x \in X \mid B_\epsilon(y) \cap F(x) \neq \emptyset\}$ is open in X, where $B_\epsilon(y) := \{y' \in \mathbf{R}^m \mid \|y - y'\| < \epsilon\}$ and the open cover $\{U(y)\}_{y \in \mathbf{R}^m}$ of X has a finite subcover. Study a partition of unity subordinate to the finite subcover.)
    (ii) Further assume that F is closed-valued. Prove that F admits a continuous selection. (*Hint*: Construct inductively a sequence

of continuous functions $\{f^q\}_q$ and a sequence of correspondences $\{F^q\}_q$ such that

$$F^q(x) := F(x) \cap B_{2^{-(q-1)}}(f^{q-1}(x)),$$

$$\rho(\{f^q(x)\}, F^q(x)) < 2^{-q}.)$$

(iii) Further assume that X is a convex subset of $\mathbf{R}^m$ and $\forall x \in X: F(x) \subset X$. Using Brouwer's fixed-point theorem and the preceding result (ii), prove that there exists a fixed-point of F.

3. Use Exercise 2(i) to prove Kakutani's fixed-point theorem. (*Hint*: Let C be a nonempty, convex, compact subset of $\mathbf{R}^n$, and let $F: C \to C$ be an u.s.c., non-empty-, closed-, and convex-valued correspondence. For each $\epsilon > 0$, define a correspondence $F^\epsilon: C \to C$ by

$$F^\epsilon(x) := \operatorname{co} \bigcup_{x' \in C \cap B_\epsilon(x)} F(x').)$$

4

Noncooperative Behavior and Equilibrium

The central concept of this chapter is the Nash equilibrium of a game in normal form. It is a solution of a game that is based on the postulate that each player behaves noncooperatively (individualistically) and passively. Its existence theorem, due to Nash, is presented as Theorem 4.1.1, and its nonoptimality is discussed in an example in Section 4.2. Debreu's generalization of a game in normal form and of a Nash equilibrium is presented in Section 4.3: an abstract economy and a social equilibrium. When the game is generalized to this extent, one can apply the results to analysis of the competitive equilibrium of a pure exchange economy. Arrow and Debreu did this, indeed; their work is presented in Section 4.4: The competitive equilibrium existence theorem (Theorem 4.4.2) is proved by applying the social equilibrium existence theorem (Theorem 4.3.1). Positive results on the optimality of competitive equilibrium, now called the fundamental theorems of welfare economics, are proved in Section 4.5 (Theorems 4.5.1 and 4.5.2). Finally, Schmeidler's recent pioneering work is presented in Section 4.6; it applies the Nash equilibrium concept in an attempt to answer an important open question that has a long history: *Which* of the given economic agents sets the competitive price vector and *in what way*? A brief bibliographical note is found in Section 4.7.

4.1. Nash Equilibrium of a Game in Normal Form‡

Given the set N of players, a game in normal form specifies a strategy set and a utility function for each player. For player j denote by X^j the set of all strategies available to him, and define $X := \prod_{j \in N} X^j$. It is postulated that his preference relation is represented by a utility function $u^j: X \times X^j \to \mathbf{R}$. The intended interpretation is that when all of the players have chosen their strategies $x := (x^1, \ldots, x^n) \in X$, player j will enjoy his utility level $u^j(x, \xi^j)$ when he changes his strategy from x^j to $\xi^j \in X^j$. Here player j passively expects the others not to change their strategies. Another (more static) interpretation is that the jth player's utility level is $u^j(x, \xi^j)$ when he chooses ξ^j and all of the others choose $(x^1, \ldots, x^{j-1}, x^{j+1}, \ldots, x^n)$, in which case u^j is assumed to be independent of x^j. Thus player i influences player j through j's utility function u^j, and *vice versa*. A *game in normal form* is now defined as a list of specified data $\{X^j, u^j\}_{j \in N}$.

The nature of a game in normal form constrained by the second interpretation, may be best understood if it is reformulated as a game in strategic form. Here a strategy bundle gives rise to an outcome $f(x)$ in the society, and it is on this set of outcomes that individual preferences are defined. More precisely, let Y be the set of possible outcomes of the society, let $v^j: Y \to \mathbf{R}$ be a utility function of player j, and let $f: X \to Y$ be an outcome function. A *game in strategic form* is a list of specified data $(\{X^j\}_{j \in N}, Y, f, \{v^j\}_{j \in N})$. Given a game in strategic form, a game in normal form is constructed by setting

$$u^j(x, \xi^j) := v^j(f(x^1, \ldots, x^{j-1}, \xi^j, x^{j+1}, \ldots, x^n)).$$

On the other hand, given a game in normal form, constrained by the second interpretation, a game in strategic form is constructed by setting $Y := X$, $f :=$ the identity function, and $v^j(y) := u^j(y, y^j)$.

‡ Part of the material in this and following sections in the chapter is based on selected ideas from Ichiishi, T., Non-Cooperation and cooperation. In M. Deistler, E. Fürst, and G. Schwödiauer (Eds.), *Games, Economic Dynamics, and Time Series Analysis*. Vienna/Würzburg, Physica-Verlag, 1982. Copyright Physica-Verlag Ges.m.b.H., Wien (Vienna) 1982.

4.1. Nash Equilibrium of a Game in Normal Form

Within the present framework one can postulate cooperative behavior of the players and hence the associated solution concept. The framework is, however, too general to yield clear-cut theorems on cooperative behavior. In this generality a type of noncooperative behavior has been extensively studied; it is summarized in the following solution concept. A *Nash equilibrium* of a game in normal form $\{X^j, u^j\}_{j \in N}$ is an n-tuple of strategies $x^* := (x^{1*}, \ldots, x^{n*}) \in X$ such that for every $j \in N$, $u^j(x^*, x^{j*}) \geq u^j(x^*, \xi^j)$ for all $\xi^j \in X^j$. The real number $u^j(x^*, x^{j*})$ is the utility level that player j currently enjoys, and in equilibrium there is no incentive for him to change his strategy x^{j*} by himself.

Theorem 4.1.1. *Let $\{X^j, u^j\}_{j \in N}$ be a game in normal form. Assume for every $j \in N$, that X^j is a nonempty, convex, compact subset of a Euclidean space, that u^j is continuous in $X \times X^j$, and that $u^j(x, \cdot)$ is quasi-concave in X^j given any $x \in X$. Then there exists a Nash equilibrium of the game.*

A complete proof of a more general theorem (Theorem 4.3.1) will be given in Section 4.3.

It would be useful to look at a special case that is customarily called a *bimatrix game*. Suppose that $n = 2$ and that X^1 and X^2 are finite say $X^j = \{1, \ldots, m_j\}$. Suppose that $u^j((x^1, x^2), \xi^j)$ does not depend on x^j (see the second interpretation given in the first paragraph of this section), and define $a_{pq} := u^1((p', q), p)$, $b_{pq} := u^2((p, q'), q)$ for $p, p' \in X^1$ and $q, q' \in X^2$. Then the game $\{X^j, u^j\}_{j=1,2}$ reduces to the bimatrix game

$$\begin{bmatrix} (a_{11}, b_{11}) & \cdots & (a_{1m_2}, b_{1m_2}) \\ \vdots & & \vdots \\ (a_{m_1 1}, b_{m_1 1}) & \cdots & (a_{m_1 m_2}, b_{m_1 m_2}) \end{bmatrix},$$

in which the first player's (the second player's, resp.) payoff is a_{pq} (b_{pq}, resp.) when the first chooses a pure strategy $p \in X^1$ and the second chooses a pure strategy $q \in X^2$, and the Nash equilibrium reduces to the *equilibrium* of the bimatrix game. In general, equilibria with pure strategies may not exist because the strategy sets X^1 and X^2 are not convex.

The preceding bimatrix is used to describe another game in normal form, in which a player can choose a mixed strategy: Let

$$\mathbf{X}^j := \{x^j \in \mathbf{R}_+^{m_j} \mid \sum_{r=1}^{m_j} x_r^j = 1\}, \quad j = 1, 2.$$

A point in the simplex $\mathbf{X}^j$ is interpreted as a mixed strategy, i.e., a probability on the pure strategies $\{1, \ldots, m_j\}$. When the two players choose mixed strategies $(x^1, x^2) \in \mathbf{X}^1 \times \mathbf{X}^2$, the expected payoff of player 1 (of player 2, resp.) is $\sum_{p,q} a_{pq} x_p^1 x_q^2$ ($\sum_{p,q} b_{pq} x_p^1 x_q^2$, resp.). Define

$$u^1(x, \xi^1) := \sum_{p=1}^{m_1} \sum_{q=1}^{m_2} a_{pq} \xi_p^1 x_q^2,$$

$$u^2(x, \xi^2) := \sum_{p=1}^{m_1} \sum_{q=1}^{m_2} b_{pq} x_p^1 \xi_q^2.$$

Then $\{\mathbf{X}^j, u^j\}_{j=1,2}$ is a game in normal form, and its Nash equilibrium exists because all the assumptions of Theorem 4.1.1 are satisfied; in particular the $\mathbf{X}^j$ are convex.

4.2. Optimality

The Nash equilibrium of a game in normal form is a *descriptive concept* in the sense that it describes what the result would be when n players play the game (given, of course, the postulate on the behavioral pattern of the players). On the other hand, one can define a *normative concept* that specifies a criterion that the result of the game is desired to satisfy. Usually, a normative concept is defined independently of any behavioral pattern of the players and is used as a target of the planner (the outside third person who acts as a regulator).

The following is a typical normative concept: Given a game in normal form $\{X^j, u^j\}_{j \in N}$, a strategy bundle $x^* \in X$ is called *Pareto optimal* if it is not true that there exists $y \in X$ such that $u^j(y, y^j) > u^j(x^*, x^{j*})$ for every $j \in N$.

4.2. Optimality

The following example of a bimatrix game, given by Albert W. Tucker, serves as an excellent counterexample in order to assert that a Nash equilibrium is not necessarily Pareto optimal:

$$\begin{bmatrix} (10, 10) & (0, 15) \\ (15, 0) & (5, 5) \end{bmatrix}.$$

The scenario for this game is the following: Suppose two players are prisoners who have been charged with some joint crime. For each player the first strategy is to plead innocent, and the second strategy is to plead guilty. If both plead guilty, they will be punished. If both plead innocent, there is no solid evidence of their crime, and so they will be punished lightly. If one pleads guilty and the other pleads innocent, the one who confesses will be pardoned for his honesty, and the one who denies wrongdoing will be punished most severely. (Caution: The numbers a_{pq} and b_{pq} in the bimatrix *are not* the terms of imprisonment but *are* the utility levels.) This game is customarily called the *prisoner's dilemma*. It is easy to check that the unique Nash equilibrium is (second row, second column) $\in X^1 \times X^2$, which is not Pareto optimal.

Even when the strategy sets are extended to the sets of mixed strategies $\mathbf{X}^1$ and $\mathbf{X}^2$, the strategy bundle (probability 1 on the second row, probability 1 on the second column) is still the unique Nash equilibrium. To see this, let $(x^{1*}, x^{2*}) \in \mathbf{X}^1 \times \mathbf{X}^2$ be a Nash equilibrium. Given x^{2*}, x_1^{1*} solves

$$\text{maximize:}_{x_1} \quad x_1 x_1^{2*} \cdot 10 + x_1 x_2^{2*} \cdot 0$$

$$+ (1 - x_1)x_1^{2*} \cdot 15 + (1 - x_1)x_2^{2*} \cdot 5,$$

subject to: $\quad 0 \leq x_1 \leq 1$.

The objective function is $-5x_1 + \{10x_1^{2*} + 5\}$. Thus $x_1^{1*} = 0$, similarly $x_1^{2*} = 0$.

To sum up, just going by the assumptions of the Nash equilibrium existence theorem (Theorem 4.1.1), one cannot hope for the existence of a Pareto optimal Nash equilibrium.

4.3. Social Equilibrium of an Abstract Economy

An abstract economy is formulated by introducing the concept of "feasibility" to a game in normal form. Let N be the set of players, let X^j be the strategy set for player j, and define $X := \prod_{j \in N} X^j$. The feasible strategy correspondence of player j is a correspondence F^j: $X \to X^j$. When all of the players have chosen their strategies $x \in X$, the feasibility of the jth player's strategies is restricted to the subset $F^j(x)$ of X^j. It is postulated that his preference relation is represented by a utility function u^j: gr $F^j \to \mathbf{R}$. Thus the other players influence player j in two ways: (1) indirectly by restricting j's feasible strategies to $F^j(x)$ and (2) directly by affecting j's utility level u^j. An *abstract economy* is a list of specified data $\{X^j, F^j, u^j\}_{j \in N}$. It should be emphasized, however, that an abstract economy is *not* a game, in spite of its mathematical generality: No player can individually play this "game," since player j must know the others' strategies in order to know his own feasible strategy set $F^j(x)$, but the others cannot determine their feasible strategies without knowing j's strategy. Thus an abstract economy is a *pseudo-game* and it is useful only as a mathematical tool to establish existence theorems in various applied contexts (see Section 4.4).

A noncooperative solution concept, Nash equilibrium of a game in normal form, can readily be extended in the present general setup. A *social equilibrium* of an abstract economy $\{X^j, F^j, u^j\}_{j \in N}$ is an n-tuple of strategies $x^* := (x^{1*}, \ldots, x^{n*}) \in X$ such that for every $j \in N$, (1) $x^{j*} \in F^j(x^*)$ and (2) $u^j(x^*, x^{j*}) \geq u^j(x^*, \xi^j)$ for all $\xi^j \in F^j(x^*)$. Condition (1) says that x^{j*} is feasible, and condition (2) says that player j cannot find a feasible strategy to bring about a higher utility level than the current level $u^j(x^*, x^{j*})$. Again player j passively expects the others to keep their strategies $(x^{1*}, \ldots, x^{j-1*}, x^{j+1*}, \ldots, x^{n*})$.

Theorem 4.3.1. *Let $\{X^j, F^j, u^j\}_{j \in N}$ be an abstract economy. Assume for every $j \in N$, that X^j is a nonempty, convex, compact subset of a Euclidean space, that F^j is both u.s.c. and l.s.c. in X, that $F^j(x)$ is nonempty, closed and convex for every $x \in X$, that u^j is continuous in gr F^j, and that $u^j(x, \cdot)$ is quasi-concave in $F^j(x)$ given any $x \in X$. Then there exists a social equilibrium of the abstract economy.*

PROOF. Given $j \in N$ and $x \in X$, denote by $\Phi^j(x)$ the set of solutions of the following maximization problem: Maximize $u^j(x, \xi^j)$ subject to $\xi^j \in F^j(x)$. All of the conditions for the maximum theorem (Theorem 2.3.1) are satisfied, and so the correspondence $\Phi^j: X \to X^j$, $x \mapsto \Phi^j(x)$, is u.s.c. For each $x \in X$, $\Phi^j(x)$ is clearly nonempty and closed. To show that it is convex choose any $\xi^{j(0)}, \xi^{j(1)} \in \Phi^j(x)$, any $t \in [0, 1]$, and define $\xi^{j(t)} := (1-t)\xi^{j(0)} + t\xi^{j(1)}$. Convexity of $F^j(x)$ implies $\xi^{j(t)} \in F^j(x)$. By quasi-concavity of $u^j(x, \cdot)$, $u^j(x, \xi^{j(t)}) \geq \min[u^j(x, \xi^{j(0)}), u^j(x, \xi^{j(1)})]$. Therefore $\xi^{j(t)} \in \Phi^j(x)$. Now define a correspondence $\Phi: X \to X$, by $\Phi(x) := \Phi^1(x) \times \cdots \times \Phi^n(x)$. The correspondence Φ satisfies all of the conditions for Kakutani's fixed-point theorem (Theorem 3.2.2). A fixed-point of Φ is easily shown to be a social equilibrium of the abstract economy. □

Note that Theorem 4.1.1 is included in Theorem 4.3.1: Just set $F^j(x) \equiv X^j$ for all $x \in X$.

4.4. Competitive Equilibrium of a Pure Exchange Economy

The social equilibrium existence theorem (Theorem 4.3.1) serves as a mathematical tool to prove the existence of equilibria for a wide class of economic models in which noncooperative behavior of the agents is postulated. The purpose of this section is to show how Theorem 4.3.1 can be applied to prove the existence of a competitive equilibrium of a pure exchange economy $\{X^i, \lesssim_i, \omega^i\}_{i=1}^m$ (formulated in Section 0.2). For each consumer i denote by $\gamma^i: \Delta^L \times \mathbf{R} \to X^i$ his budget set correspondence: $\gamma^i(p, w^i) := \{x^i \in X^i \mid p \cdot x^i \leq w^i\}$. A hint of the proof of Lemma 4.4.1 is given in Exercise 1 of Chapter 2.

Lemma 4.4.1. *Let X^i be the consumption set of consumer i, and let $(p, w^i) \in \Delta^L \times \mathbf{R}$ be a price–wealth pair. Assume that X^i is a convex, compact subset of $\mathbf{R}^l$ and that $\min\{p \cdot x^i \mid x^i \in X^i\} < w^i$. Then γ^i is both u.s.c. and l.s.c. at (p, w^i).*

Theorem 4.4.2.‡ Let $\mathscr{E} := \{X^i, \precsim_i, \omega^i\}_{i=1}^m$ be a pure exchange economy. Assume for every i, that X^i is a nonempty, convex, compact subset of $\mathbf{R}^l$, that $\precsim_i$ is complete, transitive, closed, weakly convex, and that $\min\{p \cdot x^i \mid x^i \in X^i\} < p \cdot \omega^i$ for each $p \in \Delta^L$. Then there exists a competitive equilibrium of $\mathscr{E}$.

PROOF. First, an abstract economy is constructed from $\mathscr{E}$. Let $n = m + 1$. The ith consumer is now called the ith player ($1 \leq i \leq m$), and the additional player (the nth player) is called a "market participant." For i, $1 \leq i \leq m$, his consumption set X^i is also his strategy set. Define $X^n := \Delta^L$. The market participant controls the price vector. By Theorem 0.2.1, $\precsim_i$ is represented by a continuous utility function $v^i: X^i \to \mathbf{R}$. Weak convexity of $\precsim_i$ is equivalent to quasi-concavity of v^i. Given $((x^i)_i, p) \in \prod_{j \in N} X^j$, define $F^j((x^i)_i, p) := \gamma^j(p, p \cdot \omega^j)$ for every $j = 1, \ldots, m$; $F^n((x^i)_i, p) := \Delta^L$; $u^j(((x^i)_i, p), \xi^j) := v^j(\xi^j)$ for every $j = 1, \ldots, m$; $u^n(((x^i)_i, p), \xi^n) := \xi^n \cdot \sum_{i=1}^m (x^i - \omega^i)$. By Theorems 0.2.1 and 4.4.1, the abstract economy $\{X^j, F^j, u^j\}_{j \in N}$ satisfies all of the assumptions in Theorem 4.3.1, and therefore has a social equilibrium $((x^{i*})_i, p^*) \in \prod_{i=1}^m X^i \times \Delta^L$. It is straightforward to see from the social equilibrium condition that x^{i*} is a maximal element of $\gamma^i(p^*, p^* \cdot \omega^i)$ with respect to $\precsim_i$ for every $i = 1, \ldots, m$. It is also clear that $p^* \cdot \sum_{i=1}^m (x^{i*} - \omega^i) \leq 0$. Thus since $u^n(((x^{i*})_i, p^*), p) \leq u^n(((x^{i*})_i, p^*), p^*)$ for every $p \in \Delta^L$, one concludes that $p \cdot \sum_{i=1}^m (x^{i*} - \omega^i) \leq 0$ for every $p \in \Delta^L$. In particular, the inequality is true for

$$p = (0, \ldots, \underbrace{1}_{h}, \ldots, 0),$$

so $\sum_{i=1}^m (x_h^{i*} - \omega_h^i) \leq 0$. This is true $\forall h \in L$. Thus $\sum_{i=1}^m (x^{i*} - \omega^i) \leq \mathbf{0}$. □

4.5. Fundamental Theorems of Welfare Economics

It has already been observed that some game in normal form has a unique Nash equilibrium that is not Pareto optimal (e.g., the prisoner's

---

‡ This theorem is based on ideas found in Kakutani, S., A generalization of Brouwer's fixed-point theorem. *Duke Mathematical Journal* (1941) **8**, 457–459. Copyright 1941, Duke University Press (Durham, N.C.).

4.5. Fundamental Theorems of Welfare Economics

dilemma). Positive results have, however, been established for the competitive equilibrium of a pure exchange economy (or more generally, of a private ownership economy). Recall that a pure exchange economy is a list of specified data $\mathscr{E} = \{X^i, \lesssim_i, \omega^i\}_{i=1}^m$. Since Pareto optimality of an allocation of commodity bundles (to be defined below) does not depend on the distribution $\{\omega^i\}_i$ of the initial endowment among the consumers, but depends merely on the total endowment $\sum_{i=1}^m \omega^i$ of the society, one introduces the notion of an *economy* $\mathscr{E}' := (\{X^i, \lesssim_i\}_{i=1}^m, \omega)$ where $\omega \in \mathbf{R}^l$ is the total endowment vector. An allocation of commodity bundles $(\bar{x}^i)_{i=1}^m$ is called *Pareto optimal* in an economy $\mathscr{E}'$ if (1) $\bar{x}^i \in X^i$ for every i and $\sum_{i=1}^m \bar{x}^i \leq \omega$, and if (2) it is not true that there exists $(x^i)_{i=1}^m \in \prod_{i=1}^m X^i$ such that $\sum_{i=1}^m x^i \leq \omega$ and $x^i >_i \bar{x}^i$ for every i.

Theorem 4.5.1 (First Fundamental Theorem of Welfare Economics). *Let $((x^{i*})_i, p^*)$ be a competitive equilibrium of a pure exchange economy $\{X^i, \lesssim_i, \omega^i\}_{i=1}^m$. Then $(x^{i*})_i$ is Pareto optimal in the economy $(\{X^i, \lesssim_i\}_{i=1}^m, \sum_{i=1}^m \omega^i)$.*

PROOF. If $(x^{i*})_i$ is not Pareto optimal, there exists an attainable allocation $(x^i)_i$ such that $x^i >_i x^{i*}$ for every i. For every i, $x^i \notin \gamma^i(p^*, p^* \cdot \omega^i)$, so $p^* \cdot x^i > p^* \omega^i$. Therefore $p^* \cdot \sum_i x^i > p^* \cdot \sum_i \omega^i$. On the other hand, attainability of $(x^i)_i$ and nonnegativeness of p^* imply $p^* \cdot \sum_i x^i \leq p^* \cdot \sum_i \omega^i$ —a contradiction. □

Theorem 4.5.2 (Second Fundamental Theorem of Welfare Economics). *Let $(\bar{x}^i)_i$ be a Pareto optimum of an economy $(\{X^i, \lesssim_i\}_{i=1}^m, \omega)$. Assume for each i, that X^i is a convex subset of $\mathbf{R}^l$, that $\lesssim_i$ is complete, transitive, closed, and convex, that $\bar{x}^i$ is a nonsatiation point, and that $\inf\{p \cdot x^i \mid x^i \in X^i\} < p \cdot \bar{x}^i$ for each $p \in \mathbf{R}^l \setminus \{0\}$. Then there exist $(\omega^i)_i$ that are elements of $\prod_{i=1}^m X^i$ and $\bar{p} \in \mathbf{R}^l \setminus \{0\}$ such that $\sum_{i=1}^m \omega^i \leq \omega$ and $((\bar{x}^i)_i, \bar{p})$ is a competitive equilibrium of the pure exchange economy $\{X^i, \lesssim_i, \omega^i\}_{i=1}^m$.*

PROOF. *Step 1.* For each i denote by P^i the set $\{x \in X^i \mid x >_i \bar{x}^i\}$. It is clear that P^i is nonempty, and (by the convexity of $\lesssim_i$) that $\bar{x}^i \in \bar{P}^i \setminus P^i$. It will be shown that P^i is convex. Choose any $x, x' \in P^i$, and assume that there exists $x'' \in [x, x']$ such that $x'' \lesssim_i \bar{x}^i$. By the closedness of $\lesssim_i$, the set $\{\xi \in [x, x'] \mid \xi \lesssim_i \bar{x}^i\}$ is compact. Therefore one may assume w.l.o.g. that $[x, x''[\subset P^i$. There exists $x''' \in [x, x''[$ such that $x''' <_i x'$; otherwise $x'' \gtrsim_i x'$ by the closedness of $\lesssim_i$, which contradicts the choice of x''. Then $x'' \in]x''', x'[$, and by the convexity of $\lesssim_i$, $x'' >_i x'''$ — a contradiction.

Step 2. The set $\sum_{i=1}^m P^i$ is nonempty, open and convex, and Pareto optimality of $(\bar{x}^i)_i$ implies $\sum_{i=1}^m \bar{x}^i \notin \sum_{i=1}^m P^i$. Thus by the support theorem, there exists $\bar{p} \in \mathbf{R}^l \setminus \{0\}$ such that $\bar{p} \cdot x > \bar{p} \cdot \sum_i \bar{x}^i$ for all $x \in \sum_{i=1}^m P^i$. It will be shown that for each i, $\bar{p} \cdot x^i \geq \bar{p} \cdot \bar{x}^i$ for all $x^i \in P^i$. Suppose the contrary; suppose there exist i_0, $x^{i_0} \in P^{i_0}$ such that $\bar{p} \cdot x^{i_0} < \bar{p} \cdot \bar{x}^{i_0}$. For each $i \neq i_0$, one can choose $x^i \in P^i$ sufficiently close to $\bar{x}^i$ that

$$\bar{p} \cdot x^i - \bar{p} \cdot \bar{x}^i < (m-1)^{-1}(\bar{p} \cdot \bar{x}^{i_0} - \bar{p} \cdot x^{i_0}).$$

Then

$$\bar{p} \cdot \sum_{i \neq i_0} x^i < \bar{p} \cdot \sum_{i \neq i_0} \bar{x}^i + \bar{p} \cdot \bar{x}^{i_0} - \bar{p} \cdot x^{i_0},$$

which contradicts the preceding consequence of the support theorem.

Step 3. Set $\omega^i := \bar{x}^i$. It will be shown that $x^i \in \gamma^i(\bar{p}, \bar{p} \cdot \omega^i)$ implies $x^i \lesssim_i \bar{x}^i$. Suppose the contrary; suppose there exists $x^i \in \gamma^i(\bar{p}, \bar{p} \cdot \omega^i)$ such that $x^i >_i \bar{x}^i$. Choose $b^i \in X^i$ for which $\bar{p} \cdot b^i < \bar{p} \cdot \bar{x}^i$. Since $\lesssim_i$ is closed and convex, there exists $x^{i'} \in [b^i, x^i[$ such that $x^{i'} >_i \bar{x}^i$. But $\bar{p} \cdot x^{i'} < \bar{p} \cdot \bar{x}^i$ — a contradiction of Step 2. □

The two fundamental theorems have the following economic significance. On the one hand, the price-taking behavior of consumers results in an efficient allocation of resources. On the other hand, to achieve an efficient allocation of resources, say $(\bar{x}^i)_{i=1}^m$, all the government has to do is to allocate the wealth level $\bar{p} \cdot \bar{x}^i$ to consumer i, $i = 1, \ldots, m$, and let the market mechanism work. A competitive equilibrium is then established, in which the commodity bundle $\bar{x}^i$ is precisely demanded by consumer i.

4.6. Game-Theoretical Interpretation of the Competitive Equilibrium

While the significance (in particular, the Pareto efficiency) of the price mechanism has been emphasized by economists over the past several centuries, virtually no convincing work has ever been done for explaining *which* of the given economic agents sets the price vector and *in what way*. The market participant introduced in the proof of Theorem 4.4.2, for example, is a hypothetical agent designed for the existence proof and should not be confused with any of the given m consumers. Besides, the game-theoretical interpretation of the competitive equilibrium in the proof of Theorem 4.4.2 is not based on a game (such as a game in normal form), but merely on a pseudo-game (i.e., the abstract economy). One would, therefore, hope for an economic or game-theoretical model that formulates a price-setting process in addition to the consumer behavior in the market.

Recently, David Schmeidler proposed a game in normal form, in which the choice of a price vector is a part of the strategy choice for each player and established that the Nash equilibrium of the game is precisely the competitive equilibrium. The purpose of this section is to provide a brief account of this result.

Let $\mathscr{E} := \{X^i, \lesssim_i, \omega^i\}_{i=1}^m$ be a pure exchange economy. A game in normal form is now constructed from $\mathscr{E}$. For consumer i, his strategy set is

$$S^i := \{(x^i, p^i) \in \mathbf{R}^l \times (\text{icr } \Delta^L) \mid p^i \cdot x^i = p^i \cdot \omega^i\}.$$

Assume that $X^i = \mathbf{R}^l$ and that his preference relation $\lesssim_i$ is represented by a utility function $v^i: \mathbf{R}^l \to \mathbf{R}$. Set $S := \prod_{i=1}^m S^i$. An *outcome function* $f: S \to \mathbf{R}^{lm}$, $s \mapsto (f^i(s))_{i=1}^m$, with $f^i(s) \in \mathbf{R}^l$, is defined as follows: Given a strategy bundle $s := (s^i)_i$, consumer i first finds the set of consumers who propose the same price vector as he does,

$$M_i(s) := \{j \in \{1, \ldots, m\} \mid p^j = p^i\},$$

where $s^j := (x^j, p^j)$. Exchange of commodities then takes place among the members of $M_i(s)$, and consumer i receives the commodity bundle

$$f^i(s) := x^i - \sum_{j \in M_i(s)} (x^j - \omega^j)/\#M_i(s)$$

at the end. The utility function $\tilde{u}^i: S \to \mathbf{R}$ for this game is therefore defined as

$$\tilde{u}^i(s) := v^i(f^i(s)).$$

Here the second (more static) interpretation of the game (see the first paragraph of Section 4.1) is adopted.

Theorem 4.6.1. *Let $\mathscr{E} := \{X^i, \lesssim_i, \omega^i\}_{i=1}^m$ be a pure exchange economy. Assume for every i, that $X^i = \mathbf{R}^l$, that $\lesssim_i$ is strongly convex, monotone, and representable by a differentiable utility function v^i and that the demand set $\xi^i(p) := \{x \in \gamma^i(p, p \cdot \omega^i) \mid \forall x' \in \gamma^i(p, p \cdot \omega^i): x' \lesssim_i x\}$ is nonempty for every $p \in \mathrm{icr}\,\Delta^L$. Also assume that $m \geq 3$. Let $\{S^i, \tilde{u}^i\}_{i=1}^m$ be the game in normal form, with the outcome function f, as constructed in the preceding paragraph. Then*

(i) *If s^* is a Nash equilibrium of the game $\{S^i, \tilde{u}^i\}_{i=1}^m$, then $p^{i*} = p^{j*}$ $(=:p^*$, say) for any i, j, and $((f^i(s^*))_i, p^*)$ is a competitive equilibrium of the pure exchange economy $\mathscr{E}$.*

(ii) *If $((x^{i*})_i, p^*)$ is a competitive equilibrium of $\mathscr{E}$, then there exists a Nash equilibrium s^* of $\{S^i, \tilde{u}^i\}_{i=1}^m$ for which $(f^i(s^*), p^{i*}) = (x^{i*}, p^*)$ for every i.*

PROOF. (i) Let s^* be a Nash equilibrium of $\{S^i, \tilde{u}^i\}_{i=1}^m$.

Step 1. For any distinct i, j, $f^i(s^*) \gtrsim_i \xi^i(p^{j*})$, and if $j \in M_i(s^*)$, then $f^i(s^*) = \xi^i(p^{j*})$. Indeed, if $j \in M_i(s^*)$ (i.e., if $p^{i*} = p^{j*}$), then player i could receive the outcome $\xi^i(p^{j*})$ by choosing his strategy (x^i, p^{j*}), where

$$x^i = \frac{1}{\#M_i(s^*) - 1} \left\{ \xi^i(p^{j*}) \cdot \#M_i(s^*) + \sum_{k \in M_i(s^*) \setminus \{i\}} x^{k*} - \sum_{k \in M_i(s^*)} \omega^k \right\}.$$

Since (x^{i*}, p^{j*}) is the best noncooperative strategy for i, it follows that $f^i(s^*) \gtrsim_i \xi^i(p^{j*})$. But $\xi^i(p^{j*})$ is the unique best commodity bundle in the budget set $\gamma^i(p^{j*}, p^{j*} \cdot \omega^i)$, and the fact $[f^k(s^*) \gtrsim_k \xi^k(p^{j*})$ for all $k \in M_i(s^*)$. $\sum_{k \in M_i(s^*)} f^k(s^*) = \sum_{k \in M_i(s^*)} \omega^k]$ implies $[f^k(s^*) \in \gamma^k(p^{j*}, p^{j*} \cdot \omega^k)$ for all

$k \in M_i(s^*)$]. So $f^i(s^*) = \xi^i(p^{j*})$. If $j \notin M_i(s^*)$, then player i could still receive the outcome $\xi^i(p^{j*})$ by choosing his strategy (x^i, p^{j*}), where

$$x^i = \frac{1}{\#M_j(s^*)} \{\xi^i(p^{j*}) \cdot (\#M_j(s^*) + 1) + \sum_{k \in M_j(s^*)} x^{k*} - \sum_{k \in M_j(s^*) \cup \{i\}} \omega^k\}.$$

So $f^i(s^*) \gtrsim_i \xi^i(p^{j*})$.

Step 2. If there exists i for which $\#M_i(s^*) \geq 2$, then $M_j(s^*) = \{1, \ldots, m\}$ for every j. To show this suppose the contrary; i.e., suppose there exists $j \notin M_i(s^*)$. Then $p^{i*} \neq p^{j*}$. Since $\sum_{k \in M_i(s^*)} f^k(s^*) = \sum_{k \in M_i(s^*)} \omega^k$, there are two cases:

(A) There exists $k \in M_i(s^*)$ such that $p^{j*} \cdot f^k(s^*) < p^{j*} \cdot \omega^k$; or else,
(B) For any $k \in M_i(s^*)$ it follows that $p^{j*} \cdot f^k(s^*) = p^{j*} \cdot \omega^k$.

If (A) is the case, $\xi^k(p^{j*}) >_k f^k(s^*)$ for this k by the monotonicity of $\lesssim_k$, which contradicts Step 1. If (B) is the case, $f^k(s^*) = \xi^k(p^{j*})$ for all $k \in M_i(s^*)$ since by Step 1 $f^k(s^*)$ is at least as desirable as $\xi^k(p^{j*})$, $\xi^k(p^{j*})$ is the unique best bundle in $\gamma^k(p^{j*}, p^{j*} \cdot \omega^k)$, and $f^k(s^*)$ itself is in $\gamma^k(p^{j*}, p^{j*} \cdot \omega^k)$. On the other hand, Step 1 together with $[\#M_i(s^*) \geq 2]$ implies $f^k(s^*) = \xi^k(p^{i*})$ for all $k \in M_i(s^*)$. Therefore $\xi^k(p^{i*}) = \xi^k(p^{j*})$. Since the gradient vector $(\partial v^k / \partial x_1, \ldots, \partial v^k / \partial x_l)$ of the utility function v^k at $\xi^k(p)$ is proportional to p, it follows that $p^{i*} = p^{j*}$—a contradiction.

Step 3. It suffices to show that there exists i for which $\#M_i(s^*) \geq 2$. Suppose $M_i(s^*) = \{i\}$ for every i. Then $f^i(s^*) = \omega^i$ for every i. Fix any i. There exists p uniquely in icr Δ^L such that $\xi^i(p) = \omega^i$. Since $m \geq 3$, $\#\{p^{j*} | j \neq i\} \geq 2$, so there exists $j \neq i$ such that $p^{j*} \neq p$. Then by the definition of p, $\xi^i(p^{j*}) >_i \xi^i(p)$. But by Step 1, $f^i(s^*) \gtrsim_i \xi^i(p^{j*})$, so $f^i(s^*) >_i \omega^i$—a contradiction.

(ii) Let $((x^{i*})_i, p^*)$ be a competitive equilibrium of $\mathscr{E}$. Define s^* by $s^{i*} := (x^{i*}, p^*)$ for every i. Then $f^i(s^*) = x^{i*} = \xi^i(p^*)$. So s^* is a Nash equilibrium of $\{S^i, \tilde{u}^i\}_{i=1}^m$. □

4.7. Notes

Consider the bimatrix game given in Section 4.1. When $a_{pq} + b_{pq} = 0$ for all $(p, q) \in \{1, \ldots, m_1\} \times \{1, \ldots, m_2\}$, the game is called a *zero-sum two-person game*. Here the two players have a direct conflict of interest since one's gain is equivalent to the other's loss. Let $\mathbf{X}^j := \{x^j \in \mathbf{R}_+^{m_j} \mid \sum_{r=1}^{m_j} x_r^j = 1\}$, the set of mixed strategies, for $j = 1, 2$. Von Neumann (1928) initiated game theory by establishing the *minimax principle* for this class of games: Given any $m_1 \times m_2$ matrix $((a_{pq}))$ of real numbers, it follows that

$$\max_{x^1 \in \mathbf{X}^1} \min_{x^2 \in \mathbf{X}^2} \sum_{p,q} a_{pq} x_p^1 x_q^2 = \min_{x^2 \in \mathbf{X}^2} \max_{x^1 \in \mathbf{X}^1} \sum_{p,q} a_{pq} x_p^1 x_q^2.$$

The left-hand side of this equality is the maximal security level that player 1 can achieve, and there is a corresponding interpretation of the right-hand side for player 2. This number is called the *value* of the game. Strategies (x^{1*}, x^{2*}) for which $\sum_{p,q} a_{pq} x_p^{1*} x_q^{2*}$ is the value are called *optimal strategies*. Lemma 4.7.1 (von Neumann and Morgenstern 1947, pp. 95–96) relates the minimax principle to the equilibrium existence theorem (a special case of Theorem 4.1.1).

Lemma 4.7.1. *Let X, Y be any sets, and let $f: X \times Y \to \mathbf{R}$ be a bounded function. Then $\max_{x \in X} \inf_{y \in Y} f(x, y) = \min_{y \in Y} \sup_{x \in X} f(x, y)$ if and only if there exists $(x^*, y^*) \in X \times Y$ such that for any $(x, y) \in X \times Y$, $f(x, y^*) \leq f(x^*, y^*) \leq f(x^*, y)$.*

PROOF. *Necessity.* Set

$$v := \max_x \inf_y f(x, y) = \min_y \sup_x f(x, y).$$

Choose $x^* \in X$ and $y^* \in Y$ so that

$$v = \inf_y f(x^*, y) \quad \text{and} \quad v = \sup_x f(x, y^*).$$

Then

$$v \leq f(x^*, y^*) \quad \text{and} \quad v \geq f(x^*, y^*).$$

So $v = f(x^*, y^*)$. Thus

$$f(x^*, y^*) = v = \min_y f(x^*, y) = \max_x f(x, y^*).$$

Sufficiency. For any x, $f(x, y^*) \leq f(x^*, y^*)$, so

$$\forall x : \inf_y f(x, y) \leq f(x^*, y^*). \tag{1}$$

4.7. Notes

On the other hand, for any y, $f(x^*, y^*) \leq f(x^*, y)$, which is equivalent to

$$\inf_y f(x^*, y) = f(x^*, y^*). \tag{2}$$

By conditions (1) and (2),

$$\max_x \inf_y f(x, y) = \inf_y f(x^*, y) = f(x^*, y^*).$$

Similarly, one can prove

$$\min_y \sup_x f(x, y) = f(x^*, y^*). \quad \Box$$

The pair (x^*, y^*) of Lemma 4.7.1 is called a *saddle point* of f. Now apply Lemma 4.7.1 to the preceding zero-sum two-person game

$$(X = \mathbf{X}^1, Y = \mathbf{X}^2, f(x, y) = \sum_{p,q} a_{pq} x_p y_q).$$

Then the minimax principle is equivalent to the existence of a saddle point. It is straightforward to check that a saddle point is in this case equivalent to a Nash equilibrium, whose existence is guaranteed in Theorem 4.1.1. Remark that (x^1, x^2) are optimal strategies if and only if the pair is a saddle point.

Nash (1950, 1951) introduced an n-person noncooperative game and established the special case of Theorem 4.1.1 defined by

$$X^j = \{x^j \in \mathbf{R}_+^{m_j} \mid \sum_{h=1}^{m_j} x_h^j = 1\},$$

$$u^j(x, \xi^j) = \sum_{h_1=1}^{m_1} \cdots \sum_{h_n=1}^{m_n} a_{h_1 \cdots h_n}^j x_{h_1}^1 \cdots x_{h_{j-1}}^{j-1} \xi_{h_j}^j x_{h_{j+1}}^{j+1} \cdots x_{h_n}^n,$$

given a payoff function on the pure strategies $a^j \colon \prod_{i \in N} \{1, \ldots, m_i\} \to \mathbf{R}$, $(h_1, \ldots, h_n) \mapsto a_{h_1 \cdots h_n}^j$ for every $j \in N$. The preceding paragraph showed how Nash (1950, 1951) generalized the pioneering work of von Neumann (1928). The impact of Nash (1950, 1951), however, boils down to a shift in attention from the minimax concept (such as the optimal strategies) to the equilibrium concept (e.g., the Nash equilibrium); it is only with the latter concept that game theorists could advance the

n-person noncooperative theory with full generality for $n > 2$. The equilibrium concept was actually introduced first by Cournot (1838).

Theorem 4.1.1 is based strongly on the assumption that each u^j is continuous *jointly* in x and ξ^j; otherwise, the upper semicontinuity of Φ^j (see the proof of Theorem 4.3.1) does not follow. For a partial relaxation of the joint continuity assumption see Nikaidô and Isoda (1955). Debreu (1952) formulated the abstract economy and established Theorem 4.3.1; actually, his main theorem is more general than Theorem 4.3.1. Motivated by a work of Mas-Colell (1974), Shafer and Sonnenschein (1975) established a social equilibrium existence theorem for an abstract economy in which players' preference relations may not be complete or transitive.

There are two extensions of Theorem 4.3.1 to the situation in which infinitely many players may exist. One extension (Theorem 4.7.2) is trivially included in Prakash and Sertel (1974). (Their work will be commented on in the last paragraph of this section.)

Theorem 4.7.2. *Let $\{X^a, F^a, u^a\}_{a \in A}$ be an abstract economy with the (possibly infinite) player set A. Assume for each $a \in A$, that X^a is a nonempty, convex, compact subset of a Hausdorff locally convex topological vector space, that $F^a: X \to X^a$ is both u.s.c. and l.s.c. in $X := \prod_{a \in A} X^a$ endowed with the product topology, that $F^a(x)$ is nonempty, closed, and convex for every $x \in X$, that $u^a: \mathrm{gr}\ F^a \to \mathbf{R}$ is continuous, and that $u^a(x, \cdot)$ is quasi-concave in $F^a(x)$ given any $x \in X$. Then there exists a social equilibrium of the abstract economy.*

The other extension, due to Schmeidler (1973), exploits the measure space structure of the player set. Note that in the abstract economy constructed from a given pure exchange economy (see the proof of Theorem 4.4.2), the utility function u^n of the market participant depends only on $(\sum_{i=1}^{m} x^i, \xi^n)$, rather than on $(((x^i)_{i=1}^{m}, p), \xi^n)$. Therefore the following formulation of a result given by Schmeidler (1973) will prove useful: Let $(A, \mathscr{A}, v)$ be a probability measure space of players. Besides A, there is an additional player "atom" denoted by b. Let $X(a)$ (Y, resp.) be the strategy set for player $a \in A$ (for player b, resp.), and assume that $X(a) \subset \mathbf{R}^l$. Given the correspondence $X: A \to \mathbf{R}^l, a \mapsto X(a)$, denote by $\mathscr{L}_X$ the set of integrable selections of X, i.e., the set of inte-

4.7. Notes

grable functions $f: (A, \mathscr{A}, v) \to \mathbf{R}^l$ such that $f(a) \in X(a)$, v-a.e. in A. Define $\int X := \{\int_A f\, dv \mid f \in \mathscr{L}_X\}$. The feasible strategy correspondence of player $a \in A$ (of player b, resp.) is a correspondence $F(a, \cdot, \cdot)$: $(\int X) \times Y \to X(a)$ (a correspondence $G: (\int X) \times Y \to Y$, resp.). It is postulated that the preference relation of player $a \in A$ (of player b, resp.) is represented by a utility function, $u(a, \cdot, \cdot, \cdot): (\int X) \times Y \times X(a) \to \mathbf{R}$ (by a utility function, $v: (\int X) \times Y \times Y \to \mathbf{R}$, resp.). An *abstract economy* is a list of specified data $((A, \mathscr{A}, v), \{b\}, \{X(a)\}_{a \in A}, Y, \{F(a, \cdot)\}_{a \in A}, G, \{u(a, \cdot)\}_{a \in A}, v)$. A *social equilibrium* of a given abstract economy is a pair (f^*, y^*) of members of $\mathscr{L}_X$ and Y such that (1) v-a.e. in A, $f^*(a) \in F(a, \int f^*\, dv, y^*)$ and $u(a, \int f^*\, dv, y^*, f^*(a)) \geq u(a, \int f^*\, dv, y^*, \xi^a)$ for all $\xi^a \in F(a, \int f^*\, dv, y^*)$ and (2) $y^* \in G(\int f^*\, dv, y^*)$ and $v(\int f^*\, dv, y^*, y^*) \geq v(\int f^*\, dv, y^*, \eta)$ for all $\eta \in G(\int f^*\, dv, y^*)$. A correspondence $\Gamma: (A, \mathscr{A}, v) \to \mathbf{R}^l$ is called integrably bounded if there exists an integrable function $f: (A, \mathscr{A}, v) \to \mathbf{R}_+$ such that $[x \in \Gamma(a)]$ implies $[\|x\| \leq f(a)]$, v-a.e. in A. Denote by $\mathscr{B}(\mathbf{R}^l)$ the Borel σ-algebra generated by the open subsets of $\mathbf{R}^l$.

Theorem 4.7.3. *Let* $((A, \mathscr{A}, v), \{b\}, \{X(a)\}_{a \in A}, Y, \{F(a, \cdot)\}_{a \in A}, G, \{u(a, \cdot)\}_{a \in A}, v)$ *be the abstract economy constructed above. There exists a social equilibrium of the abstract economy if the following conditions* (1)–(4′) *are true*: (1) $(A, \mathscr{A}, v)$ *is an atomless probability measure space* (2) X *is integrably bounded*, $\operatorname{gr} X \in \mathscr{A} \otimes \mathscr{B}(\mathbf{R}^l)$, *and* $X(a)$ *is nonempty and closed in* $\mathbf{R}^l$, v-a.e. *in* A. (2′) Y *is a nonempty, convex, compact subset of a Hausdorff locally convex topological vector space*. (3) $\operatorname{gr} F(\cdot, w, y) \in \mathscr{A} \otimes \mathscr{B}(\mathbf{R}^l)$ *for every* $(w, y) \in (\int X) \times Y$; $F(a, \cdot, \cdot)$ *is both u.s.c. and l.s.c. and is non-empty-valued, closed-valued*, v-a.e. *in* A. (3′) G *is both u.s.c. and l.s.c. and is non-empty-valued, closed-valued, convex-valued.* (4) $u(\cdot, w, y, \cdot): \operatorname{gr} X \to \mathbf{R}$ *is measurable for every* $(w, y) \in (\int X) \times Y$; $u(a, \cdot, \cdot, \cdot): (\int X) \times Y \times X(a) \to \mathbf{R}$ *is continuous* v-a.e. *in* A. (4′) $v: (\int X) \times Y \times Y \to \mathbf{R}$ *is continuous, and* $v(w, y, \cdot)$ *is quasi-concave in* Y *for every* $(w, y) \in (\int X) \times Y$.

Note that the convexity assumptions are dropped from the atomless part.

Arrow and Debreu (1954) established Theorem 4.4.2. Their method of proof is reproduced here; i.e., to construct an abstract economy, given a more specific economic model and then to apply Theorem 4.3.1. They also relaxed the boundedness assumption on the consumption sets X^i of Theorem 4.4.2 in the following way: Assume each consumption set X^i is closed and is bounded from below; i.e., there exists $b^i \in X^i$ such that $X^i \subset \{b^i\} + \mathbf{R}^l_+$. Then the set of attainable states

$$A(\{1, \ldots, m\}) := \{(x^i)_i \in \prod_{i=1}^{m} X^i \mid \sum_{i=1}^{m} x^i \leq \sum_{i=1}^{m} \omega^i\}$$

is compact. Define $K^{(q)} := \{x \in \mathbf{R}^l \mid \|x\| \leq q\}$, and for a given economy $\mathscr{E}$ consider the sequence of truncated economies $\{\mathscr{E}^{(q)}\}_{q=1,2,\ldots}$, defined by $\mathscr{E}^{(q)} := \{X^i \cap K^{(q)}, \precsim_i, \omega^i\}_{i=1}^{m}$. Apply Theorem 4.4.2 to $\mathscr{E}^{(q)}$ for all q sufficiently large, and obtain a competitive equilibrium $((x^{i,(q)})_{i=1}^{m}, p^{(q)})$ of $\mathscr{E}^{(q)}$. Since the sequence $\{((x^{i,(q)})_i, p^{(q)})\}_q$ of equilibria lies in a compact set $A(\{1, \ldots, m\}) \times \Delta^L$, it has a convergent subsequence. It is easy to check that the limit of the subsequence is a competitive equilibrium of $\mathscr{E}$. Thus one establishes

Theorem 4.7.4. *Let* $\mathscr{E} := \{X^i, \precsim_i, \omega^i\}_{i=1}^{m}$ *be a pure exchange economy. Under the same assumptions as those in Theorem 4.4.2, except that* [X^i *is bounded*] *is relaxed to* [X^i *is bounded from below*], *there exists a competitive equilibrium of* $\mathscr{E}$.

Further work was done on the existence of competitive equilibria during the 1950s, in particular by Gale (1955), Nikaidô (1956), Debreu (1956), and McKenzie (1959): This work is unified in Debreu (1959). Negishi (1960) supplied an existence proof, in which he looked at only the set of Pareto optimal allocations and chose from the set a particular allocation that is proved to be a competitive allocation. Mas-Colell (1974) strengthened the competitive equilibrium existence theorem by dropping the completeness and transitivity assumptions on the preference relations.

Recall that the convexity assumption on $\precsim_i$ and the convexity of X^i guarantee convex-valuedness of the individual demand correspondences. The approach of Gale (1955) and Nikaidô (1956) was to look at the *total* demand correspondence and apply Kakutani's fixed-point theorem. Now if the economy consists of "many" agents, then the total demand correspondence becomes "almost" convex-valued even if the individual preference relations are not convex (so that the

4.7. Notes

individual demand correspondences are not convex-valued). This is clearly a consequence of the Shapley–Folkman theorem (Theorem 1.6.5). Starr (1969) was the first to have this insight precisely, and he established the existence of approximate equilibria for a large economy with nonconvex preference relations.

Aumann (1964) formulated a model of pure exchange with an atomless measure space of consumers and then subsequently established a competitive equilibrium existence theorem (Aumann, 1966). Let $(A, \mathscr{A}, v)$ be an atomless probability measure space of consumers. Each consumer a is characterized by a triple $(X(a), \lesssim_a, \omega(a))$ of his consumption set (a subset of $\mathbf{R}^l$), his preference relation on $X(a)$, and his initial endowment (a point of $\mathbf{R}^l$). The function $\omega: (A, \mathscr{A}, v) \to \mathbf{R}^l$, $a \mapsto \omega(a)$, is assumed to be integrable. A *competitive equilibrium* is a pair (f^*, p^*) of members of $\mathscr{L}_X$ and Δ^L such that (1) $f^*(a)$ is a maximal element of $\{x \in X(a) \mid p^* \cdot x \leq p^* \cdot \omega(a)\}$ with respect to $\lesssim_a$, v-a.e. in A and (2) $\int f^* \, dv \leq \int \omega \, dv$. Now introduce a fictitious "market participant"; call him player b. Just as Arrow and Debreu (1954) used Theorem 4.3.1 to prove Theorem 4.4.2, one can use Theorem 4.7.3 to supply an alternative proof of the existence theorem of Aumann (1966). Note that the (weak) convexity assumption on $\lesssim_a$ is dropped here. The convexity assumption on $X(a)$ is not explicit in Theorem 4.7.3, but it is usually used to establish lower semicontinuity of the budget set correspondence $p \mapsto \{x \in X(a) \mid p \cdot x \leq p \cdot \omega(a)\}$ (see Exercise 1 (vii) of Chapter 2). Yamazaki (1978) dropped the convexity assumption on $X(a)$ and supplied a sufficient condition under which the budget set correspondence is l.s.c., v-a.e. in A.

The fundamental theorems of welfare economics had been discussed by many theorists during the 1930s; the clearest treatment of this issue using the calculus can be found in Lange (1942). Arrow (1951) boiled down the logical essence of the issue to the separation principle. Section 4.5 is based on Arrow's insight.

Section 4.6 is based on Schmeidler (1980), and Exercise 1 is based on Uzawa (1962). Exercise 2 is the simplest version of the issue pioneered by Debreu (1970) and developed under the title "Regular Differentiable Economies."

Some other work related to noncooperative solution concepts is noted: (1) Moulin (1979) proposed the concept of *reversible point* for a two-person game in normal form and characterized such points. The reversible points are strategy bundles such that there exists an (infinitesimal) move in strategy, that is best when each player behaves individualistically, but is bad from a cooperative point of view. He also extended his analysis to a three-person game. (2) Prakash and Sertel (1974) generalized the abstract economy to the situation in which the utility and also the feasible strategy set of a player depend not only on the others' choices of strategy but also on the others' feasible strategy sets (e.g., on the enemy's capabilities). Their logic for the existence theorem goes deep (beyond fixed-point theorems for topological vector spaces).

EXERCISES

1. Denote by $\Delta^L := \{p \in \mathbf{R}^l \mid p \geq \mathbf{0}, \sum_{h=1}^{l} p_h = 1\}$ the price domain. A *total excess demand function* is a function $z: \Delta^L \to \mathbf{R}^l$ such that the Walras law

$$\forall p \in \Delta^L : p \cdot z(p) = 0$$

holds true. A *competitive equilibrium* is a price vector $p^* \in \Delta^L$ such that

$$z(p^*) \leq \mathbf{0} \quad \text{and} \quad z_h(p^*) = 0 \quad \text{if} \quad p_h^* > 0.$$

Equilibrium Existence Theorem. Let z be a continuous total excess demand function. Then there exists a competitive equilibrium.

Brouwer's Fixed-Point Theorem. Let f be a continuous function from Δ^L to Δ^L. Then there exists a fixed-point of f.

(i) Use Brouwer's fixed-point theorem to prove the equilibrium existence theorem. (*Hint*: Given a total excess demand function z, consider a function $f: \Delta^L \to \Delta^L$ defined by

$$f_h(p) := \frac{p_h + \max[0, z_h(p)]}{1 + \sum_{k=1}^{l} \max[0, z_k(p)]}.)$$

(ii) Use the equilibrium existence theorem to prove Brouwer's fixed-point theorem. (*Hint*: Given a function $f: \Delta^L \to \Delta^L$, construct a total excess demand function $z: \Delta^L \to \mathbf{R}^l$,

$$z_h(p) := f_h(p) - p_h \cdot \frac{p \cdot f(p)}{p \cdot p}.)$$

2. Let Z be the set of all C^1 total excess demand functions

$$Z := \left\{ z: \Delta^L \to \mathbf{R}^l \,\middle|\, \begin{array}{l} z \text{ is continuously differentiable in an} \\ \text{open neighborhood of } \Delta^L \text{ in aff } \Delta^L \\ \forall p \in \Delta^L : p \cdot z(p) = 0. \end{array} \right\}.$$

The set Z is endowed with a metric d defined by

$$d(z, z') := \max_{p \in \Delta^L} \|z(p) - z'(p)\| + \max_{p \in \Delta^L} \sum_{h=1}^{l} \left\| \frac{\partial}{\partial p_h} z(p) - \frac{\partial}{\partial p_h} z'(p) \right\|.$$

A member of Z is also called an economy. An economy z is called *regular* if

$$(\forall p \in \Delta^L : z(p) = \mathbf{0}) : Dz(p) \text{ is of rank } (l-1).$$

(i) Prove that for every regular economy, the number of equilibria*, $\#\{p \in \Delta^L \mid z(p) = \mathbf{0}\}$ is finite. (*Hint*: the inverse function theorem and compactness of Δ^L.)

(ii) Denote by $\mathcal{R}$ the set of all regular economies. Prove that $\mathcal{R}$ is open and dense in (Z, d). Conclude that it is a generic property of an economy to have a finite set of equilibria. (*Hint*: One can use the following special case of Thom's transversal density theorem: For each $\epsilon > 0$ and each set of $(l-1)$ C^1 functions $f_1, \ldots, f_{l-1}$ from Δ^L to $\mathbf{R}$, there exists a set of $(l-1)$ C^1 functions $g_1, \ldots, g_{l-1}$ from Δ^L to $\mathbf{R}$ such that

$$|f_h(p) - g_h(p)| + \sum_{k=1}^{l} \left| \frac{\partial}{\partial p_k} f_h(p) - \frac{\partial}{\partial p_k} g_h(p) \right| < \epsilon$$

---

* The definition of equilibrium in Problem 2 is somewhat different from the definition given in Problem 1.

for all $p \in \Delta^L$, $h = 1, \ldots, l - 1$, and at each $p \in \Delta^L$ for which $g_h(p) = 0$ for all h,

$$\det \begin{bmatrix} \dfrac{\partial g_1}{\partial p_1} & \cdots & \dfrac{\partial g_1}{\partial p_{l-1}} \\ \vdots & & \vdots \\ \dfrac{\partial g_{l-1}}{\partial p_1} & \cdots & \dfrac{\partial g_{l-1}}{\partial p_{l-1}} \end{bmatrix} \neq 0.)$$

5

Cooperative Behavior and Stability

The central concepts of this chapter are (1) the core of a game in characteristic function form, with or without side-payments, and (2) the social coalitional equilibrium of the society. Two types of a game in characteristic function form are presented: a side-payment game and a non-side-payment game. The former is a special case of the latter; both may be considered special cases of a game in normal form. A cooperative solution concept that characterizes stability, the core, is defined for each type, and a theorem for its nonemptiness is proved (Bondareva–Shapley's theorem (Theorem 5.2.1) for nonemptiness of the core of a side-payment game, and Scarf's theorem (Theorem 5.4.1) for nonemptiness of the core of a non-side-payment game). Section 5.1 (Section 5.3, resp.) supplies a mathematical tool essentially useful for the proof of Theorem 5.2.1 (Theorem 5.4.1, resp.). The framework of a non-side-payment game is general enough to formulate a cooperative behavior of the economic agents in a pure exchange economy. Given a pure exchange economy, a commodity allocation that gives rise to a point in the core is called a core allocation. Scarf established a core allocation existence theorem (Theorem 5.5.2) as a straightforward application of his theorem (Theorem 5.4.1). Section 5.6 presents Anderson's contribution to the Edgeworth conjecture that establishes a close relationship between competitive allocation (a noncooperative solution concept) and core allocation (a cooperative solution concept) within the framework of a pure exchange economy. One leaves the economic model and returns to game theory in Section 5.7. The most general game-theoretical model, called a society, is constructed. A solution concept for the society, called a social coalitional equilibrium, is presented, and a theorem for its existence is proved (Theorem 5.7.1). The social coalitional equilibrium is considered a synthesis of the social equilibrium of an abstract economy (a typical noncooperative solution concept; see Section 4.3) and the core of a non-side-payment game (a typical cooperative solution concept; see Section 5.4). Theorem 5.7.1 is applied in Section 5.8 to answer a question left open in Section 4.2: conditions for the existence of a Pareto optimal Nash equilibrium. A brief bibliographical note is found in Section 5.9.

5.1. Linear Inequalities

Let A be an $m \times n$ matrix and A_j be its jth column, and let b be an $m \times 1$ matrix. Consider the following system of linear equalities with a nonnegativity constraint:

$$Ax = b, \quad x \geq 0. \qquad [*]$$

The system $[*]$ is called *consistent* if there exists $x \in \mathbf{R}^n$ satisfying condition $[*]$; such a vector x is called a *solution* of $[*]$. Given the matrix A and column vector b, one can consider a variety of linear (in)equality systems other than $[*]$ and ask about their consistency. For any system a characterization of its consistency is obtained by an application of the separation principle. Actually, once a characterization of consistency of one system, say of $[*]$, is obtained, one can use it to provide quite easily a characterization of consistency of any other system.

Lemma 5.1.1 (Minkowski–Farkas). *Let A be an $m \times n$ matrix, and let b be an $m \times 1$ matrix. The system*

$$Ax = b, \quad x \geq 0, \qquad [*]$$

5.1. Linear Inequalities

is consistent iff for every $1 \times m$ matrix λ for which $\lambda A \geq \mathbf{0}$, it follows that $\lambda b \geq 0$.

PROOF.   *Step 1.*   Define a subset $K(A)$ of $\mathbf{R}^m$ by

$$K(A) := \{x_1 A_1 + \cdots + x_n A_n \mid x_j \geq 0, j = 1, \ldots, n\}.$$

Clearly, $K(A)$ is convex and contains $\mathbf{0}$. It will be shown that $K(A)$ is closed in $\mathbf{R}^m$. Choose any sequence $\{y^q\}_q$ in $K(A)$ that converges to $\bar{y}$ in $\mathbf{R}^m$. By the proof of Lemma 1.1.3, there exists $F^q \subset \{1, \ldots, n\}$ such that $\{A_j \mid j \in F^q\}$ is linearly independent, and y^q is a positive linear combination of $\{A_j \mid j \in F^q\}$, say

$$y^q = \sum_{j \in F^q} x_j^q A_j, \qquad x_j^q > 0.$$

Passing through a subsequence if necessary, one may assume w.l.o.g. that $F^q = F$ for all q. The sequence $\{y^q\}_q$ and its limit $\bar{y}$ are then in the subspace $S := \text{span}\{A_j \mid j \in F\}$. Set $\bar{y} = \sum_{j \in F} \bar{x}_j A_j$. Since $\{A_j \mid j \in F\}$ is a basis for S, $[y^q \to \bar{y}]$ implies $[x_j^q \to \bar{x}_j$, for each $j \in F]$. Therefore $\bar{x}_j \geq 0$ for every $j \in F$; i.e., $\bar{y} \in K(A)$.

Step 2.   The system $[*]$ is consistent iff $b \in K(A)$. If $b \in K(A)$, it is easy to show that for every $1 \times m$ matrix λ for which $\lambda A \geq \mathbf{0}$ it follows that $\lambda b \geq 0$. If $b \notin K(A)$, one can strictly separate $K(A)$ and $\{b\}$ by a hyperplane in $\mathbf{R}^m$; there exist $\lambda \in \mathbf{R}^m \setminus \{\mathbf{0}\}$ and $t \in \mathbf{R}$ such that $\lambda y \geq t$ for all $y \in K(A)$ and such that $\lambda b < t$. On the one hand, $t \leq 0$ since $\mathbf{0} \in K(A)$. On the other hand, $\lambda y \geq 0$ for all $y \in K(A)$; for if there exists $y^0 \in K(A)$ such that $\lambda y^0 < 0$, then for $r > 0$ sufficiently large $\lambda(ry^0) < t$ which contradicts the above separation result because $ry^0 \in K(A)$. Thus one may assume w.l.o.g. $t = 0$; so for every $j = 1, \ldots, n$, $\lambda A e^j \geq 0$, i.e., $\lambda A \geq \mathbf{0}$.   □

Corollary 5.1.2.   *Let A be an $m \times n$ matrix, and let b be an $m \times 1$ matrix. The system*

$$Ax \geq b \qquad [**]$$

is consistent iff for every $1 \times m$ matrix λ for which $\lambda \geq \mathbf{0}$ and $\lambda A = \mathbf{0}$, it follows that $\lambda b \leq 0$.

PROOF. The system $[**]$ is consistent iff the system

$$[A, -A, -I] \begin{bmatrix} x^1 \\ x^2 \\ y \end{bmatrix} = b, \quad \begin{bmatrix} x^1 \\ x^2 \\ y \end{bmatrix} \geq \mathbf{0},$$

is consistent, where I is the $m \times m$ unit matrix. By the Minkowski–Farkas lemma, this is the case iff for every $1 \times m$ matrix λ for which $\lambda(A, -A, -I) \geq \mathbf{0}$, it follows that $\lambda b \geq 0$. □

5.2. Core of a Side-Payment Game

Let N be given as the set of players, and hence let $\mathcal{N}(:= 2^N \setminus \{\emptyset\})$ be given as the set of nonempty coalitions of players. When the players of a coalition S jointly choose the strategy bundle $(x^j)_{j \in S}$, and each member j in S makes a commitment to such a decision, one says that cooperation among the members of S is made or, equivalently, that the coalition S forms. A cooperative game explicitly formulates what each coalition can achieve by the cooperation of its members. The simplest game of this class would be a *game in characteristic function form with side payments* (or simply, a *side-payment game*) defined as a function $v: \mathcal{N} \to \mathbf{R}$. The intended interpretation is that $v(S)$ is the maximal total payoff of the coalition S that the members of S can bring about by their cooperation regardless of the actions of the players outside S.

A side-payment game v is a special case of a game in normal form. Indeed, given a side-payment game v, the associated game in normal form can be constructed as follows: For each player j, the strategy set X^j is the union of mutually disjoint sets $X^{j,S}$ with respect to the S for which $S \ni j$. If $x^j \in X^{j,S}$, then player j cooperates with the members in S. Define $u^j(x, \xi^j) := -\infty$ if for the coalition S for which $\xi^j \in X^{j,S}$, there exists $i \in S$ such that $x^i \notin X^{i,S}$. Also, let $u^j(x, \xi^j)$ be independent of $(x^i)_{i \in N \setminus S}$, where S is the coalition for which $\xi^j \in X^{j,S}$. Now the required

5.2. Core of a Side-Payment Game

game is such that

$$\{(u^j(x, x^j))_{j \in S} \mid \forall j \in S : x^j \in X^{j,S}\} = \{(u_j)_{j \in S} \mid \sum_{j \in S} u_j \leq v(S)\}$$

for every $S \in \mathcal{N}$. The associated game in normal form is, however, too cumbersome to work with. One usually discards the strategy sets and utility functions and studies directly the function v.

An important cooperative solution concept is now given: The *core of a side-payment game* v is the set $C(v)$ of all $u \in \mathbf{R}^n$ such that (1) $\sum_{j \in N} u_j \leq v(N)$ and (2) it is not true that there exists $S \in \mathcal{N}$ for which $\sum_{j \in S} u_j < v(S)$. Condition (1) is the feasibility of the payoff vector u via the grand coalition N, and condition (2) is the stability of u, i.e., no coalition can improve upon u. The core $C(v)$ is the set of solutions of the 2^n linear inequality system $-\chi_N \cdot x \geq -v(N)$, and $\chi_S \cdot x \geq v(S)$ for all $S \in \mathcal{N}$. Therefore the logic for nonemptiness of $C(v)$ is included in Corollary 5.1.2.

A subfamily $\mathcal{B}$ of $\mathcal{N}$ is called *balanced*, if there exists an indexed set $\{\lambda_S \mid S \in \mathcal{B}\}$ of nonnegative numbers such that $\sum_{S \in \mathcal{B}} \lambda_S \chi_S = \chi_N$. Note that $\sum_{S \in \mathcal{B}} \lambda_S \chi_S = \chi_N$ iff $\sum_{S \in \mathcal{B}: S \ni j} \lambda_S = 1$ for every $j \in N$. The set $\{\lambda_S \mid S \in \mathcal{B}\}$ is called the *associated balancing coefficients*. A side-payment game v is called *balanced* if for every balanced family $\mathcal{B}$ with the associated balancing coefficients $\{\lambda_S \mid S \in \mathcal{B}\}$, it follows that $\sum_{S \in \mathcal{B}} \lambda_S v(S) \leq v(N)$. The following theorem says that a side-payment game has a nonempty core iff it is balanced.

Theorem 5.2.1. *Let* $v: \mathcal{N} \to \mathbf{R}$ *be a side-payment game. The core of* v *is nonempty if and only if for every indexed set of nonnegative numbers* $\{\lambda_S \mid S \in \mathcal{N}\}$ *for which* $\sum_{S \in \mathcal{N}} \lambda_S \chi_S = \chi_N$, *it follows that* $\sum_{S \in \mathcal{N}} \lambda_S v(S) \leq v(N)$.

PROOF. By Corollary 5.1.2, $C(v) \neq \emptyset$ iff for every set $\{\lambda'_S \mid S \in \mathcal{N}\} \cup \{\mu\}$ of nonnegative numbers for which $\sum_{S \in \mathcal{N}} \lambda'_S \chi_S + \mu(-\chi_N) = \mathbf{0}$, it follows that $\sum_{S \in \mathcal{N}} \lambda'_S v(S) + \mu(-v(N)) \leq 0$. If $\mu = 0$, then $\lambda'_S = 0$ for all $S \in \mathcal{N}$ so the last inequality is automatically satisfied. If $\mu > 0$, define

$\lambda_S := \lambda'_S/\mu$, and the preceding condition reduces to the required condition of the theorem. □

5.3. K–K–M–S Theorem

Before discussing a more general cooperative game, a non-side-payment game, it would be useful to discuss Shapley's (1973) generalization of the K–K–M theorem (Theorem 3.1.2), customarily called the K–K–M–S theorem.

Theorem 5.3.1. *Let $\{C_S\}_{S \in \mathcal{N}}$ be a family of closed subsets of Δ^N, indexed by the members of $\mathcal{N}$, such that for each $T \in \mathcal{N}$, $\Delta^T \subset \bigcup_{S \subset T} C_S$. Then there exists a balanced collection $\mathcal{B}$ for which $\bigcap_{S \in \mathcal{B}} C_S \neq \emptyset$.*

PROOF. Denote by $I(x)$ the collection $\{S \in \mathcal{N} \mid C_S \ni x\}$, and define a correspondence $G: \Delta^N \to \Delta^N$ by $G(x) := \text{co}\{(1/\#S)\chi_S \mid S \in I(x)\}$. It is trivial to check that G is u.s.c. and is non-empty- and convex-valued. So is a constant correspondence F defined by $F(x) := \{(1/n)\chi_N\}$. Choose any $x \in \Delta^N$ and any $p \in \mathbf{R}^n$ such that $p \cdot x \leq p \cdot y$ for every $y \in \Delta^N$. The subset T of N for which $x \in \text{icr } \Delta^T$ is uniquely determined, and for this set T, $p \cdot x = p \cdot y$ for every $y \in \Delta^T$. By the present assumption, there exists $S \in I(x)$ such that $S \subset T$; i.e., there exists $(1/\#S)\chi_S \in G(x)$ such that $p \cdot (1/n)\chi_N \geq p \cdot (1/\#S)\chi_S$. Thus all of the conditions for the coincidence theorem (Theorem 3.3.3) are satisfied, and there exists $x^* \in \Delta^N$ such that $F(x^*) \cap G(x^*) \neq \emptyset$, i.e., $(1/n)\chi_N \in G(x^*)$. Therefore $I(x^*)$ is balanced. Needless to say $\bigcap_{S \in I(x^*)} C_S \ni x^*$. □

Note that if $C_S = \emptyset$ for every S for which $\#S \geq 2$, then Theorem 5.3.1 is reduced to the K–K–M theorem (Theorem 3.1.2).

5.4. Core of a Non-Side-Payment Game

The concept of "total payoff, $v(S)$," of the coalition S in a side-payment game v is sometimes hard to interpret when the behavior, cooperative or noncooperative, is ultimately motivated by the individual pursuit of maximal satisfaction. A more general game is presented here to overcome this difficulty. For each coalition S define

$$\mathbf{R}^S := \{x \in \mathbf{R}^n | \forall j \notin S : x_j = 0\}.$$

A *game in characteristic function form without side-payments* (or simply, a *non-side-payment game*) is a correspondence $\tilde{V} \colon \mathcal{N} \to \mathbf{R}^n$ such that $\tilde{V}(S) \subset \mathbf{R}^S$ for every $S \in \mathcal{N}$. The intended interpretation is that $u \in \tilde{V}(S)$ iff cooperation of the members in S can bring about the utility allocation $(u_i)_{i \in S}$ to the members of S. As an equivalent definition, a *non-side-payment game* is a correspondence $V \colon \mathcal{N} \to \mathbf{R}^n$ such that $[u, v \in \mathbf{R}^n, \forall i \in S : u_i = v_i]$ implies $[u \in V(S)$ if and only if $v \in V(S)]$ for every $S \in \mathcal{N}$. Clearly, $V(S)$ is a cylinder based on $\tilde{V}(S)$; i.e., $V(S) = \{u \in \mathbf{R}^n | (u_i \delta_{iS})_{i \in N} \in \tilde{V}(S)\}$, where $\delta_{iS} = 1$ if $i \in S$ and $\delta_{iS} = 0$ if $i \notin S$. The second definition of a non-side-payment game will be adopted throughout the text.

Just as in the second paragraph of Section 5.2, one can regard a non-side-payment game as a special case of a game in normal form. Again, however, it is easier to work directly with the attainable utility allocations $V(S)$.

The *core* of a non-side-payment game V is the set $C(V)$ of all $u \in \mathbf{R}^n$ such that (1) $u \in V(N)$ and (2) it is not true that there exist $S \in \mathcal{N}$ and $u' \in V(S)$ such that $u_j < u'_j$ for every $j \in S$. Condition (1) is the feasibility of the utility allocation u via the grand coalition N, and condition (2) is the stability of u.

A non-side-payment game V is called *balanced* if for every balanced subfamily $\mathscr{B}$ of $\mathcal{N}$, it follows that $\bigcap_{S \in \mathscr{B}} V(S) \subset V(N)$. Essentially the balancedness condition is sufficient (but not necessary) for nonemptiness of the core of V.

Theorem 5.4.1. *Let* $V \colon \mathcal{N} \to \mathbf{R}^n$ *be a non-side-payment game, and* define $b \in \mathbf{R}^n$ by $b_j := \sup\{u_j \in \mathbf{R} \mid u \in V(\{j\})\}$ for each $j \in N$. The core of V is nonempty if

(1) $V(S) - \mathbf{R}^n_+ = V(S)$ for every $S \in \mathcal{N}$;
(2) there exists $M \in \mathbf{R}$ such that for every $S \in \mathcal{N}$, $[u \in V(S) \cap [\{b\} + \mathbf{R}^n_+]]$ implies $[u_i < M$ for every $i \in S]$;
(3) $V(S)$ is closed in $\mathbf{R}^n$ for every $S \in \mathcal{N}$; and
(4) V is balanced.

PROOF. Without loss of generality, one may assume $b = \mathbf{0}$. Given the positive number M of the present assumption (2), define $D^S := \mathrm{co}\{-Mne^j \mid j \in S\}$ for every $S \in \mathcal{N}$. The function $\tau \colon D^N \to \mathbf{R}$, defined by $\tau(y) := \max\{r \in \mathbf{R} \mid y + r\chi_N \in \bigcup_{S \in \mathcal{N}} V(S)\}$, is continuous by the maximum theorem (Theorem 2.3.1). Therefore the function $f \colon \Delta^N \to \mathbf{R}^n$, defined by $f(y) := y + \tau(y)\chi_N$, is also continuous. Define $C_S := \{y \in D^N \mid f(y) \in V(S)\}$. Since C_S is the inverse image under f of a closed set $V(S)$, it is closed (see Figure 5.4.1). One claims that $[S, T \in \mathcal{N}, D^T \cap C_S \neq \emptyset]$ implies $[S \subset T]$. This is trivial if $T = N$. So assume that $\#T < n$, and take any $y \in D^T \cap C_S$. On the one hand, $\sum_{j \in T} y_j = -Mn$, so there exists $t \in T$ for which $y_t \leq -Mn/\#T < -M$. Since $f(y) \in \mathbf{R}^n_+$, $y_t + \tau(y) \geq 0$. Therefore $\tau(y) > M$. On the other hand, $f(y) \in V(S)$, so for every $j \in S$, $y_j + \tau(y) < M$. Thus $y_j < 0$ for every $j \in S$, from which $S \subset T$, and the claim has been proved. All of the conditions for Theorem 5.3.1 are now satisfied; there exists a balanced collection $\mathscr{B}^*$ and a point $y^* \in D^N$ such that $y^* \in \bigcap_{S \in \mathscr{B}^*} C_S$. So $f(y^*) \in \bigcap_{S \in \mathscr{B}^*} V(S) \subset V(N)$. By the definition of τ, $f(y^*)$ is on the frontier of $\bigcup_{S \in \mathcal{N}} V(S)$, and in view of assumption (1) it cannot be improved upon by any coalition. Thus $f(y^*) \in C(V)$. □

Assumption (1) of the preceding theorem is customarily called the *comprehensiveness* condition. Assumption (3) can be replaced by a weaker condition: $V(N)$ is closed in $\mathbf{R}^n$.

At this point it would be useful to introduce one more concept: A *coalition structure* is a partition of N; it describes which coalitions are formed and coexist. Not all of the coalition structures may be admissible (e.g., from the legal point of view) so denote by $\mathscr{T}$ the family of admis-

5.4. Core of a Non-Side-Payment Game

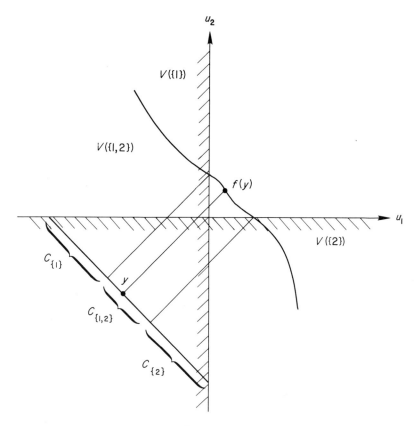

Figure 5.4.1

sible coalition structures. A *generalized non-side-payment game* is a pair $(V, \mathscr{T})$ of a non-side-payment game and a family of admissible coalition structures. The *core* of $(V, \mathscr{T})$ is the set of all pairs $(u^*, \mathscr{T}^*)$ of members of $\mathbf{R}^n$ and $\mathscr{T}$ such that (1) $u^* \in V(T)$ for every $T \in \mathscr{T}^*$ and that (2) it is not true that there exist $S \in \mathscr{N}$ and $u' \in V(S)$ such that $u_j < u'_j$ for every $j \in S$. Exactly the same argument as in the proof of Theorem 5.4.1 (except that the statement $[\bigcap_{S \in \mathscr{B}^*} V(S) \subset V(N)]$ is replaced by the statement $[\bigcap_{S \in \mathscr{B}^*} V(S) \subset \bigcup_{\mathscr{T} \in \mathscr{T}} \sum_{T \in \mathscr{T}} \tilde{V}(T)]$, where $\tilde{V}(T) := V(T) \cap \mathbf{R}^T$) establishes that

The core of a generalized non-side-payment game $(V, \mathcal{T})$, is nonempty if conditions (1)–(3) of Theorem 5.4.1 are satisfied and if game $(V, \mathcal{T})$ is balanced; i.e., for every balanced family $\mathcal{B}$, $\bigcap_{S \in \mathcal{B}} V(S) \subset \bigcup_{\mathcal{T} \in \mathcal{T}} \sum_{T \in \mathcal{T}} \tilde{V}(T)$.

Finally a sidepayment game can be considered as a special case of a non-side-payment game. Indeed, given a side-payment game v, the associated non-side-payment game V is defined by

$$V(S) := \{u \in \mathbf{R}^n | \sum_{j \in S} u_j \leq v(S)\}.$$

In this case it is straightforward to check that the core $C(v)$ of v (defined in the third paragraph of Section 5.2) and the core $C(V)$ of V (defined in the third paragraph of the present section) are identical and that conditions (1)–(3) of Theorem 5.4.1 are automatically satisfied. Call the balancedness condition on v (defined in the paragraph preceding Theorem 5.2.1) the B–Sh condition, and call the balancedness condition on V (defined in the paragraph preceding Theorem 5.4.1) the Sc condition. Then in this case, the B–Sh condition and the Sc condition are equivalent. Indeed if the Sc condition is satisfied, then $C(V) \neq \emptyset$ by Theorem 5.4.1. Since $C(V) = C(v)$, the B–Sh condition holds true by Theorem 5.2.1. On the other hand assume the B–Sh condition, choose a balanced family $\mathcal{B}$ with the associated balancing coefficients $\{\lambda_S | S \in \mathcal{B}\}$, and take any $u \in \bigcap_{S \in \mathcal{B}} V(S)$. For each $S \in \mathcal{B}$, $\sum_{j \in S} u_j \leq v(S)$. Then $\sum_{j \in N} u_j = \sum_{j \in N} (\sum_{S \in \mathcal{B}: S \ni j} \lambda_S) u_j = \sum_{S \in \mathcal{B}} \lambda_S \sum_{j \in S} u_j \leq \sum_{S \in \mathcal{B}} \lambda_S v(S) \leq v(N)$. Therefore $u \in V(N)$, and the Sc condition has been proved.

5.5. Core Allocation of a Pure Exchange Economy

Let $\mathscr{E} := \{X^j, \lesssim_j, \omega^j\}_{j \in N}$ be a pure exchange economy with n consumers (as formulated in Section 0.2, with the change that $m = n$ here), in which each preference relation $\lesssim_j$ is represented by a utility function u^j. A type of cooperative behavior of the consumers in $\mathscr{E}$ can be summarized by the associated non-side-payment game $V : \mathcal{N} \to \mathbf{R}^n$, defined by $V(S) := \{u \in \mathbf{R}^n | \exists (x^i)_{i \in S} \in \prod_{i \in S} \in X^i : \sum_{i \in S} x^i \leq \sum_{i \in S} \omega^i$, and

5.5. Core Allocation of a Pure Exchange Economy

for every $i \in S: u_i \leq u^i(x^i)\}$. A *core allocation* of $\mathscr{E}$ is an allocation that gives rise to a member of the core of the associated game V. Actually, one can give its definition without resorting to the derivative concept of utility function (hence without resorting to V): A *core allocation* of $\mathscr{E}$ is an n-tuple of commodity bundles $(x^{1*}, \ldots, x^{n*}) \in \prod_{j \in N} X^j$ such that (1) $\sum_{j \in N} x^{j*} \leq \sum_{j \in N} \omega^j$ and (2) it is not true that there exist $S \in \mathcal{N}$ and $(x^i)_{i \in S} \in \prod_{i \in S} X^i$ such that $\sum_{i \in S} x^i \leq \sum_{i \in S} \omega^i$ and $x^i >_i x^{i*}$ for every $i \in S$. The purpose of this section is to show how Theorem 5.4.1 can be applied to prove the existence of a core allocation of $\mathscr{E}$.

For the case $l = n = 2$ and $X^i = \mathbf{R}_+^2$ for $i = 1, 2$, the core allocations are illustrated in the Edgeworth box diagram (see Figure 5.5.1). By comparing Figures 0.2.1 and 5.5.1, one might conjecture that the set of competitive allocations is included in the set of core allocations. This is indeed the case, as will be proved in the next section. At the same time this fact explains the reason why the assumptions of the core allocation existence theorem (Theorem 5.5.2) are weaker than the assumptions of the competitive equilibrium existence theorem (Theorem 4.7.4).

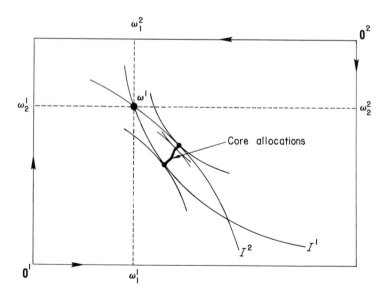

Figure 5.5.1

Lemma 5.5.1 Let $\mathscr{E} := \{X^j, \precsim_j, \omega^j\}_{j\in N}$ be a pure exchange economy in which each preference relation $\precsim_j$ is represented by a utility function $u^j\colon X^j \to \mathbf{R}$. If X^j is a convex subset of $\mathbf{R}^l$, $\omega^j \in X^j$, and $\precsim_j$ is weakly convex for every $j \in N$, then the associated game V is balanced.

PROOF. Let $\mathscr{B}$ be a balanced family with the associated balancing coefficients $\{\lambda_S | S \in \mathscr{B}\}$, and choose any $u \in \bigcap_{S\in\mathscr{B}} V(S)$. Then for each $S \in \mathscr{B}$, there exists $(x^{i,S})_{i\in S} \in \prod_{i\in S} X^i$ such that $\sum_{i\in S} x^{i,S} \leq \sum_{i\in S} \omega^i$ and $u_i \leq u^i(x^{i,S})$ for every $i \in S$. Define $\bar{x}^j := \sum_{S\in\mathscr{B}: S\ni j} \lambda_S x^{j,S}$ for every $j \in N$. By convexity of X^j, $\bar{x}^j \in X^j$; and by quasi-concavity of u^j, $u_j \leq u^j(\bar{x}^j)$. Moreover

$$\sum_{j\in N} \bar{x}^j = \sum_{j\in N} \sum_{S\in\mathscr{B}: S\ni j} \lambda_S x^{j,S} = \sum_{S\in\mathscr{B}} \lambda_S \sum_{j\in S} x^{j,S}$$

$$\leq \sum_{S\in\mathscr{B}} \lambda_S \sum_{j\in S} \omega^j = \sum_{j\in N} \sum_{S\in\mathscr{B}: S\ni j} \lambda_S \omega^j = \sum_{j\in N} \omega^j.$$

Thus $u \in V(N)$. □

Denote by $A(S)$ the set of attainable states for the coalition S:

$$A(S) := \left\{(x^i)_{i\in S} \in \prod_{i\in S} X^i \,\Big|\, \sum_{i\in S} x^i \leq \sum_{i\in S} \omega^i \right\}.$$

It is clear that the set $A(S)$ is bounded if for every $i \in S$, X^i is bounded from below, i.e., if there exists $b^i \in \mathbf{R}^l$ such that $X^i \subset \{b^i\} + \mathbf{R}^l_+$.

Theorem 5.5.2. Let $\mathscr{E} := \{X^j, \precsim_j, \omega^j\}_{j\in N}$ be a pure exchange economy. If X^j is a convex, closed subset of $\mathbf{R}^l$ and is bounded from below, $\omega^j \in X^j$, and if $\precsim_j$ is complete, transitive, closed, and weakly convex for each $j \in N$, then there exists a core allocation of $\mathscr{E}$.

PROOF. Let $u^j\colon X^j \to \mathbf{R}$ be a continuous utility function that represents $\precsim_j$ (see Theorem 0.2.1). It will be shown that all four assumptions in Theorem 5.4.1 are satisfied for the associated game V. Condition (1) (comprehensiveness) is a consequence of the definition of $V(S)$. To show conditions (2) and (3), it suffices to note that $\tilde{V}(S) = \{\tilde{u}^S(A(S))\} - \mathbf{R}^S_+$, where the continuous function $\tilde{u}^S\colon \prod_{j\in S} X^j \to \mathbf{R}^S$ is defined by

$\tilde{u}_j^S((x^i)_{i\in S}) := u^j(x^j)$ if $j \in S$ and by $\tilde{u}_j^S((x^i)_{i\in S}) := 0$ if $j \notin S$, and that $A(S)$ is compact. Condition (4) (balancedness) is the result of Lemma 5.5.1. □

5.6. A Limit Theorem of Cores

The purpose of this section is to present a work of Anderson (1978) on the *Edgeworth proposition* that in a "large" economy the set of competitive allocations is characterized as the set of core allocations.

Theorem 5.6.1. *Let* $((x^{i*})_{i=1}^m, p^*)$ *be a competitive equilibrium of a pure exchange economy* $\mathscr{E} := \{X^i, \lesssim_i, \omega^i\}_{i=1}^m$. *Then* $(x^{i*})_i$ *is a core allocation of* $\mathscr{E}$.

PROOF. Since $\sum_i x^{i*} \leq \sum_i \omega^i$, it suffices to show that the allocation cannot be improved upon by any coalition. Suppose the contrary; there exist a nonempty coalition S and $(x^i)_i \in \prod_{i\in S} X^i$ such that (1) $\sum_{i\in S} x^i \leq \sum_{i\in S} \omega^i$ and (2) $x^i >_i x^{i*}$ for every $i \in S$. Since x^{i*} is a maximal element of $\gamma^i(p^*, p^* \cdot \omega^i)$ with respect to $\lesssim_i$, condition (2) implies $p^* \cdot x^i > p^* \cdot \omega^i$ for every $i \in S$; hence $\sum_{i\in S} p^* \cdot x^i > \sum_{i\in S} p^* \cdot \omega^i$. On the other hand, nonnegativity of p^* and condition (1) imply $p^* \cdot \sum_{i\in S} x^i \leq p^* \cdot \sum_{i\in S} \omega^i$—a contradiction. □

The involved part of the Edgeworth proposition is the "converse" of Theorem 5.6.1 for a "large" economy. A type of the "converse" may be illustrated as follows. Consider the Edgeworth box diagram for the economy $\mathscr{E} := \{\mathbf{R}_+^2, \lesssim_i, \omega^i\}_{i=1}^2$. The *q-replica* of $\mathscr{E}$ is the economy $\mathscr{E}^{(q)}$ with $2q$ consumers in which there are q agents having consumer characteristics $(\mathbf{R}_+^2, \lesssim_i, \omega^i)$ for each i. Under mild assumptions, one can easily establish the *equal treatment property* of a core allocation of $\mathscr{E}^{(q)}$: each of the q agents of the same type holds the same vector as his commodity bundle (Exercise 7). A core allocation of $\mathscr{E}^{(q)}$ is therefore represented by a point in the Edgeworth box diagram; the point specifies the commodity bundle of a *representative agent* of each type (see Figure 5.6.1). Let C represent a core allocation when $q = 1$, which is not a competitive allocation. Then the tangent line of the indifference curve I^i at C does not pass through ω^1; in other words, on the line passing through both C and ω^1, one can choose points A and B such that $A >_1 C$ and $(\omega - B) >_2 (\omega - C)$, where $\omega := \omega^1 + \omega^2$. Points A and B can be chosen so that there exist integers m_1, m_2 for which

$$m_1(A - \omega^1) = m_2(B - \omega^1).$$

For any q-replica economy, with $q \geq \max[m_1, m_2]$, one can choose a coalition S that consists of m_1 agents of type 1 and m_2 agents of type 2. It is now proved that the given allocation (represented by point C) is improved on by coalition S. Assign commodity bundle A to each agent of type 1 in S and commodity bundle $(\omega - B)$ to each agent of type 2 in S. It suffices to show that this commodity allocation within S is feasible. But

$$m_1 A + m_2(\omega - B) = m_1(\omega^1 + (A - \omega^1)) + m_2(\omega^2 + (\omega^1 - B))$$
$$= (m_1 \omega^1 + m_2 \omega^2) + (m_1(A - \omega^1) - m_2(B - \omega^1))$$
$$= m_1 \omega^1 + m_2 \omega^2.$$

Thus the commodity allocation represented by point C cannot be a core allocation of $\mathscr{E}^{(q)}$ for any $q \geq \max[m_1, m_2]$.

One now moves on to the general treatment of the required "converse" due to Anderson (1978). The setup here is more general than that of replica economies. Recall that the simplex Δ^L is chosen to be the price domain (Section 0.2) and that a strict preference relation $>_i$ may be used (instead of a utility function u^i) in defining the core allocation (Section 5.5).

Theorem 5.6.2. *Let $\mathscr{E} := \{\mathbf{R}_+^l, >_i, \omega^i\}_{i \in N}$ be a pure exchange economy with a finite consumer set N. Assume for each $i \in N$, that the strict preference relation $>_i$ satisfies*

(1) *weak monotonicity:* $[x, y \in \mathbf{R}_+^l, x \gg y]$ *implies* $[x >_i y]$ *and*
(2) *free disposal:* $[x, y, z \in \mathbf{R}_+^l, x \gg y >_i z]$ *implies* $[x >_i z]$,

5.6. A Limit Theorem of Cores

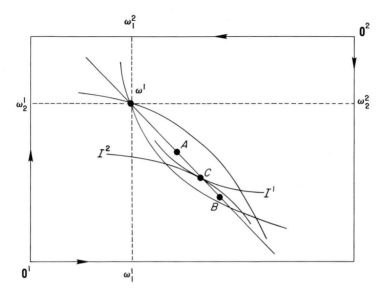

Figure 5.6.1

and that $\omega^i \in \mathbf{R}_+^l$. Set

$$M := \max\{\sum_{i \in N'} \omega_h^i \,|\, h \in \{1, \ldots, l\}.\ N' \subset N.\ \#N' = l\}.$$

Then for each core allocation $(x^{i*})_{i \in N}$ of $\mathscr{E}$ there exists $p^* \in \Delta^L$ such that

(i) $(1/\#N) \sum_{i \in N} |p^* \cdot (x^{i*} - \omega^i)| \leq 2M/\#N$ and
(ii) $(1/\#N) \sum_{i \in N} |\inf\{p^* \cdot (x - \omega^i) \,|\, x >_i x^{i*}\}| \leq 2M/\#N$.

Notice that assumptions (1) and (2) on the preference relations here are substantially weaker than those given in Section 0.2. In particular, neither closedness nor weak convexity is needed, and the set of assumptions (monotonicity and transitivity) is replaced by the weaker set of assumptions (weak monotonicity and free disposal). If the size of the economy is large (i.e., if $\#N$ is large), so that one may safely assume that $2M/\#N$ is small, conclusion (i) asserts that x^{i*} is on the average close to the budget surface $\{x \in \mathbf{R}_+^l \,|\, p^* \cdot x = p^* \cdot \omega^i\}$. Also, for $2M/\#N$ sufficiently small, conclusion (ii) asserts that minimal expenditure in the preferred set $\{x \in \mathbf{R}_+^l \,|\, x >_i x^{i*}\}$ is on the average close to the wealth level $p^* \cdot \omega^i$ or, equivalently, that x^{i*} is on the average preferred to any commodity bundle x with $p^* \cdot x < p^* \cdot \omega^i - 2M/\#N$. Thus the given core allocation $(x^{i*})_{i \in N}$, together with the price vector p^*, becomes an "approximate" competitive equilibrium of $\mathscr{E}$. The content of the term "approximation" can be made sharper if, in addition, the preference relations are strictly convex. Roughly speaking, under such stronger assumptions, one can assert that x^{i*} is close to the demand: the maximal element in the budget set $\{x \in \mathbf{R}_+^l \mid p^* \cdot x \le p^* \cdot \omega^i\}$ with respect to $>_i$; see Section 5.9.

Another way to look at the assertion of Theorem 5.6.2 is

Corollary 5.6.3. *Let* $\{\mathscr{E}^q\}_q$ *be a sequence of pure exchange economies:* $\mathscr{E}^q := \{\mathbf{R}_+^l, >_{i,q}, \omega^{i,q}\}_{i \in N_q}$. *Suppose that each* $\mathscr{E}^q$ *satisfies all of the assumptions of Theorem 5.6.2. Set* $M_q := \max\{\sum_{i \in N'} \omega_h^{i,q} \mid h \in \{1, \ldots, l\}$. $N' \subset N_q$. $\#N' = l\}$.
Assume also that $M_q/\#N_q \to 0$ *as* $q \to \infty$. *If* $(x^{i,q*})_{i \in N_q}$ *is a core allocation of* $\mathscr{E}^q$ *for each* q, *then there exists* $p^{q*} \in \Delta^L$ *for each* q *such that*

(i) $(1/\#N_q) \sum_{i \in N_q} |p^{q*} \cdot (x^{i,q*} - \omega^{i,q})| \to 0$ *and*
(ii) $(1/\#N_q) \sum_{i \in N_q} |\inf\{p^{q*} \cdot (x - \omega^{i,q}) | x >_{i,q} x^{i,q*}\}| \to 0$.

PROOF OF THEOREM 5.6.2. Recall that P^i is the preferred set $\{x \in \mathbf{R}_+^l \mid x >_i x^{i*}\}$, which is nonempty. Define $\Gamma^i := P^i - \{\omega^i\}$; and $\Phi := (1/\#N) \sum_{i \in N} (\Gamma^i \cup \{\mathbf{0}\})$.

Step 1. $(-\mathbf{R}_+^l) \cap \Phi = \varnothing$. Indeed, if there exists $z \in \Phi$ with $z \ll \mathbf{0}$, then there exist a nonempty subset S of N and $z^i \in \Gamma^i$ for each $i \in S$ such that $z = (1/\#N) \sum_{i \in S} z^i$. It will be shown that the coalition S can improve upon $(x^{i*})_i$. Define for each $i \in S$, $x^i := z^i + \omega^i - (\#N/\#S)z$. Then $x^i \gg z^i + \omega^i >_i x^{i*}$, so by the free disposal assumption (2), it follows that $x^i >_i x^{i*}$ for every $i \in S$. Moreover

$$\sum_{i \in S} x^i = \sum_{i \in S} z^i + \sum_{i \in S} \omega^i - (\#N)z = \sum_{i \in S} \omega^i.$$

Step 2. Denote by χ_L the vector in $\mathbf{R}^l$ each of whose components is 1, and define $Z := \{z \in \mathbf{R}^l \mid z \ll -(M/\#N)\chi_L\}$. Then $Z \cap \text{co } \Phi = \varnothing$. Indeed, if there exists $z \in Z \cap \text{co } \Phi$, then by the Shapley–Folkman theorem (Theorem 1.6.5) member z has a representation

$$z = (1/\#N) \sum_{i \in S} z^i + (1/\#N) \sum_{i \in T} z^i,$$

5.6. A Limit Theorem of Cores

with $S \cup T = N$ and $\#T \leq l$ such that $z^i \in (\Gamma^i \cup \{\mathbf{0}\})$ for all $i \in S$ and such that $z^i \in [\mathrm{co}(\Gamma^i \cup \{\mathbf{0}\})]\setminus(\Gamma^i \cup \{\mathbf{0}\})$ for all $i \in T$. Define $(\bar{z}^i)_{i \in N}$ by $\bar{z}^i := z^i$ if $i \in S$ by and $\bar{z}^i := \mathbf{0}$ if $i \in T$. Then $(1/\#N)\sum_{i \in N} \bar{z}^i \in \Phi$. However $\mathrm{co}(\Gamma^i \cup \{\mathbf{0}\}) \geq -\omega^i$ because $\Gamma^i \geq -\omega^i$. So

$$(1/\#N)\sum_{i \in N} \bar{z}^i = z - (1/\#N)\sum_{i \in T} z^i$$

$$\leq z + (1/\#N)\sum_{i \in T} \omega^i$$

$$\ll (1/\#N)\{-M\chi_L + \sum_{i \in T} \omega^i\}$$

$$\leq \mathbf{0}.$$

This contradicts Step 1.

Step 3. By the separation theorem (Theorem 1.5.4), there exists $p^* \in \mathbf{R}^l\setminus\{\mathbf{0}\}$ and $r \in \mathbf{R}$ such that $p^* \cdot z \geq r$ for every $z \in \mathrm{co}\,\Phi$ and $p^* \cdot z \leq r$ for every $z \in Z$. From the last result, one can easily show that $p^* \geq \mathbf{0}$. One may therefore assume w.l.o.g. that $p^* \in \Delta^L$.

Step 4. For each $\epsilon > 0$, $x^{i*} + \epsilon\chi_L - \omega^i \in \Gamma^i$, by the weak monotonicity assumption (1). Therefore

$$p^* \cdot (x^{i*} - \omega^i) \geq \inf p^* \cdot (\Gamma^i \cup \{\mathbf{0}\}) \quad \text{for every} \quad i \in N.$$

Denote by N_+ the set $\{i \in N \mid p^* \cdot (x^{i*} - \omega^i) \geq 0\}$. Then

$$(1/\#N)\sum_{i \notin N_+} p^* \cdot (x^{i*} - \omega^i)$$

$$\geq (1/\#N)\sum_{i \notin N_+} \inf p^* \cdot (\Gamma^i \cup \{\mathbf{0}\})$$

$$\geq (1/\#N)\sum_{i \notin N_+} \inf p^* \cdot (\Gamma^i \cup \{\mathbf{0}\})$$

$$+ (1/\#N)\sum_{i \in N_+} \inf p^* \cdot (\Gamma^i \cup \{\mathbf{0}\})$$

$$= (1/\#N)\sum_{i \in N} \inf p^* \cdot (\Gamma^i \cup \{\mathbf{0}\})$$

$$= \inf p^* \cdot \Phi$$

$$= \inf p^* \cdot \mathrm{co}\,\Phi \geq r \geq \sup p^* \cdot Z$$

$$= -M/\#N.$$

Since the core allocation is attainable

$$\sum_{i\notin N_+} p^* \cdot (x^{i*} - \omega^i) = - \sum_{i\in N_+} p^* \cdot (x^{i*} - \omega^i).$$

To sum up,

$$(1/\#N) \sum_{i\notin N_+} p^* \cdot (x^{i*} - \omega^i) = -(1/\#N) \sum_{i\in N_+} p^* \cdot (x^{i*} - \omega^i)$$

$$\geq \inf p^* \cdot \Phi \geq -M/\#N.$$

Now the two assertions follow easily:

$$(1/\#N) \sum_{i\in N} |p^* \cdot (x^{i*} - \omega^i)|$$

$$= (2/\#N) \sum_{i\notin N_+} |p^* \cdot (x^{i*} - \omega^i)| \leq 2M/\#N,$$

$$(1/\#N) \sum_{i\in N} |\inf\{p^* \cdot (x - \omega^i) | x >_i x^{i*}\}|$$

$$= (1/\#N) \sum_{i\in N_+} |\inf p^* \cdot \Gamma^i| + (1/\#N) \sum_{i\notin N_+} |\inf p^* \cdot \Gamma^i|$$

$$\leq (1/\#N)\{- \sum_{i\in N_+} \inf p^* \cdot (\Gamma^i \cup \{\mathbf{0}\}) + \sum_{i\in N_+} p^* \cdot (x^{i*} - \omega^i)\}$$

$$- (1/\#N) \sum_{i\notin N_+} \inf p^* \cdot (\Gamma^i \cup \{\mathbf{0}\})$$

$$= -\inf p^* \cdot \Phi + (1/\#N) \sum_{i\in N_+} p^* \cdot (x^{i*} - \omega^i)$$

$$\leq M/\#N + M/\#N. \quad \square$$

5.7. Social Coalitional Equilibrium of a Society

The two strands in game theory, presented in Chapter 4 and in this Chapter, are unified here. A more general behavioral pattern is formu-

5.7. Social Coalitional Equilibrium of a Society

lated, the relevant solution concept proposed and its existence theorem is provided.

The price-taking behavior of a consumer can be regarded as a particular case of the passive, noncooperative behavior formulated in Section 4.3 (see the proof of Theorem 4.4.2). It is undeniable that in free societies the market mechanism, which assumes price-taking behavior of economic agents, works to a large extent. On the other hand, researchers of the theory of the firm have long emphasized certain nonmarket mechanisms as essential determinants in a modern capitalistic economy. These nonmarket mechanisms involve cooperative behavior of the agents. Thus to build a descriptive model of a modern capitalistic economy, one needs a game-theoretical foundation for a behavioral pattern that contains both noncooperation and cooperation. It turns out that the theorem established in this general setup (Theorem 5.7.1) can also supply an answer to a welfare question on a Nash equilibrium of a game in normal form (see Section 4.2).

Given the set of players N, and hence given the set of nonempty coalitions of players $\mathcal{N}$, an abstract model of a society is now constructed. The model explicitly formulates how the members of a coalition are influenced by outsiders. For player j denote by X^j his strategy set, and define $X^S := \prod_{j \in S} X^j$ for every $S \in \mathcal{N}$. Put $X := X^N$ for notational convenience. The feasible strategy correspondence of the coalition S is a correspondence $F^S: X \to X^S$. The subset $F^S(x)$ of X^S is interpreted as the set of all feasible strategy bundles for S as a cooperative unit when $x := (x^1, \ldots, x^n) \in X$ has been chosen. It is postulated that the preference relation of the jth player in S, when the members of S agree to cooperate, is represented by a utility function $u_S^j: \operatorname{gr} F^S \to \mathbf{R}$. Here the dependence of the utility function on S reflects the possible fact that the members in S enjoy the environment (i.e., nonstrategical elements) specific to S. Thus the players outside a coalition S influence the members of S (1) indirectly by restricting the feasible joint strategies of the members of S to $F^S(x)$ and (2) directly by affecting the utility level u_S^j for all $j \in S$ (compare this with the first paragraph of Section 4.3). Let $\mathcal{T}$ be the set of admissible coalition structures. A *society* is now defined as a list of specified data $(\{X^j\}_{j \in N}, \{F^S\}_{S \in \mathcal{N}}, \{u_S^j\}_{j \in S \in \mathcal{N}}, \mathcal{T})$.

A *social coalitional equilibrium* of a society is a pair consisting of a strategy bundle and an admissible coalition structure, $(x^*, \mathcal{T}^*) \in X \times$

$\mathscr{T}$, such that (1) for each $T \in \mathscr{T}^*$, $x^{T*} := (x^{j*})_{j \in T} \in F^T(x^*)$ and (2) it is not true that for some $S \in \mathcal{N}$ there exists $\xi^S \in F^S(x^*)$ for which $u_S^j(x^*, \xi^S) > u_{T(j)}^j(x^*, x^{T(j)*})$ for every $j \in S$, where $T(j)$ is the member of $\mathscr{T}^*$ such that $T(j) \ni j$. Condition (1) signifies the feasibility condition. The term $u_S^j(x^*, \xi^S)$ of condition (2) is the utility level that player j would get by participating in the new coalition S and agreeing to take the strategy ξ^S. Here he passively accepts the strategies currently taken by the players outside S. The term $u_{T(j)}^j(x^*, x^{T(j)*})$ is the utility level he currently enjoys. Condition (2) signifies, therefore, the stability condition.

Theorem 5.7.1. *Let* $(\{X^j\}_{j \in N}, \{F^S\}_{S \in \mathcal{N}}, \{u_S^j\}_{j \in S \in \mathcal{N}}, \mathscr{T})$ *be a society. There exists a social coalitional equilibrium of the society if*

(1) X^j *is a nonempty, convex, compact subset of a Euclidean space for every* $j \in N$;

(2) F^S *is both u.s.c. and l.s.c. in* X, *and* $F^S(x)$ *is nonempty, closed for any* $x \in X$, *for every* $S \in \mathcal{N}$;

(3) u_S^j *is continuous in* gr F^S *for every* $S \in \mathcal{N}$ *and every* $j \in S$;

(4) *given any* $(x, v) \in X \times \mathbf{R}^n$, $[$*there exists a balanced collection* $\mathscr{B}$ *such that for each* $S \in \mathscr{B}$ *there exists* $\xi(S) \in F^S(x)$ *for which* $v_j \leq u_S^j(x, \xi(S))$ *for every* $j \in S]$ *implies* $[$*there exist* $\mathscr{T} \in \mathscr{T}$ *and* $\eta^T \in F^T(x)$ *for every* $T \in \mathscr{T}$ *such that* $v_j \leq u_T^j(x, \eta^T)$ *for all* $j \in T \in \mathscr{T}]$; *and if*

(5) *given any* $(x, v) \in X \times \mathbf{R}^n$, *the set* $\bigcup_{\mathscr{T} \in \mathscr{T}} \{\xi \in X \mid$ *for every* $T \in \mathscr{T}, \xi^T \in F^T(x),$ *and* $v_j \leq u_T^j(x, \xi^T)$ *for each* $j \in T\}$ *is convex.*

In order to restate assumption (4) of Theorem 5.7.1 in somewhat more familiar terms, define for a given $x \in X$ a non-side-payment game $V_x: \mathcal{N} \to \mathbf{R}^n$ by $V_x(S) := \{v \in \mathbf{R}^n \mid \exists \xi^S \in F^S(x): \forall j \in S: v_j \leq u_S^j(x, \xi^S)\}$, for each $S \in \mathcal{N}$. Then assumption (4) says that the generalized non-side-payment game $(V_x, \mathscr{T})$ is balanced (see the second-to-last paragraph of Section 5.4) for every $x \in X$. When $\mathscr{T}$ is a singleton $\{\mathscr{T}\}$, assumption (5) of Theorem 5.7.1 is reduced to convexity of

$$\prod_{T \in \mathscr{T}} \{\xi^T \in X^T \mid \xi^T \in F^T(x), v_j \leq u_T^j(x, \xi^T) \text{ for each } j \in T\}$$

for every $(x, v) \in X \times \mathbf{R}^n$, and under compactness of X^j and continuity of u_T^j, the latter condition is equivalent to [convexity of $F^T(x)$ and

5.7. Social Coalitional Equilibrium of a Society

quasi-concavity of $u_T^j(x, \cdot)$ on $F^T(x)$, for every $x \in X$ and $j \in T$, given any $T \in \mathcal{T}$]. Thus assumption (5) may be regarded as an extension of this last convexity assumption to the case in which $\#\mathcal{T}$ can be greater than 1.

PROOF. *Step 1.* Define

$$\tilde{V}_x(S) := V_x(S) \cap \mathbf{R}^S, \quad H_x := \bigcup_{\mathcal{T} \in \mathscr{T}} \sum_{T \in \mathcal{T}} \tilde{V}_x(T).$$

By assumptions (1)–(3) and by the finiteness of the cardinality of $\mathcal{T}$, it is straightforward to check that the correspondence $H: X \to \mathbf{R}^n$, $x \mapsto H_x$, is both u.s.c. and l.s.c. The correspondence $V(S): X \to \mathbf{R}^n$, $x \mapsto V_x(S)$ is also u.s.c. and l.s.c. for each $S \in \mathcal{N}$.

Step 2. Define $m(x) \in \mathbf{R}^n$ by $m_j(x) := \max\{u_{\{j\}}^j(x, \xi^j) \mid \xi^j \in F^{\{j\}}(x)\}$. Then there exists a real number $M > 0$ such that for every $x \in X$, every $S \in \mathcal{N}$, and every $v \in V_x(S)$, $v_j - m_j(x) < M$ for all $j \in S$. Denote by D^S the set $\text{co}\{-Mne^j \mid j \in S\}$. Put $\chi'_S := -Mn\chi_S/(\#S)$; this is a member of D^S. Define a function $\tau: X \times D^N \to \mathbf{R}$ by

$$\tau(x, y) := \max\{r \in \mathbf{R} \mid y + m(x) + r \cdot \chi_N \in \bigcup_{S \in \mathcal{N}} V_x(S)\}.$$

The function τ is well-defined and is continuous by Step 1. The function $f: X \times D^N \to \mathbf{R}^n$, $(x, y) \mapsto y + m(x) + \tau(x, y) \cdot \chi_N$ is also continuous. Finally, define $C_S(x) := \{y \in D^N \mid f(x, y) \in V_x(S)\}$. For each $S \in \mathcal{N}$, the correspondence $C_S: X \to D^N$, $x \mapsto C_S(x)$ is u.s.c. Now by the same argument as that in the claim of the proof of Theorem 5.4.1, one can show that for every $x \in X$, $[S, T \in \mathcal{N}, D^T \cap C_S(x) \neq \emptyset]$ implies $[S \subset T]$.

Step 3. Define a correspondence $F: X \times D^N \to X \times D^N$ by

$$F(x, y) := \text{co} \bigcup_{\mathcal{T} \in \mathscr{T}} \{\xi \in X \mid \forall T \in \mathcal{T}: \xi^T \in F^T(x), \text{ and}$$

$$\forall w \in H_x: \|f(x, y) - u(x, \xi)\| \leq \|f(x, y) - w\|\} \times \{\chi'_N\},$$

where the jth component of $u(x, \xi)$ is $u_{T(j)}^j(x, \xi^{T(j)})$ with $j \in T(j) \in \mathcal{T}$. Define another correspondence $G: X \times D^N \to X \times D^N$ by

$$G(x, y) := \{x\} \times \text{co}\{\chi'_S \mid C_S(x) \ni y\}.$$

Both F and G are u.s.c. in $X \times D^N$ and are non-empty- and convex-valued. Choose any $(x, y) \in X \times D^N$ and any $(p, q) \in \mathbf{R}^{\dim X} \times \mathbf{R}^n$ such that $(p, q) \cdot (x, y) \leq (p, q) \cdot (x', y')$ for every $(x', y') \in X \times D^N$. By the same argument as in the proof of Theorem 5.3.1, there exist $(\xi, \chi'_N) \in F(x, y)$ and $(x, \chi'_S) \in G(x, y)$ such that $(p, q) \cdot (\xi, \chi'_N) \geq (p, q) \cdot (x, \chi'_S)$.

Step 4. By Step 3, all of the conditions for the coincidence theorem (Theorem 3.3.3) are satisfied: There exist $(x^*, y^*), (\bar{x}, \bar{y}) \in X \times D^N$ such that $(\bar{x}, \bar{y}) \in F(x^*, y^*) \cap G(x^*, y^*)$. This means that

(i) $\bar{x}$ is an element of co $\bigcup_{\mathcal{T} \in \mathcal{T}} \{\xi \in X | \forall T \in \mathcal{T} : \xi^T \in F^T(x^*)$, and $\forall w \in H_{x*} : \| f(x^*, y^*) - u(x^*, \xi) \| \leq \| f(x^*, y^*) - w \| \}$;

(ii) $\bar{y} = \chi'_N$;

(iii) $\bar{x} = x^*$; and

(iv) $\bar{y} \in \text{co}\{\chi'_S | C_S(x^*) \ni y^*\}$.

By (ii) and (iv), the collection $\mathcal{B}^* := \{S \in \mathcal{N} | C_S(x^*) \ni y^*\}$ is balanced: Thus $f(x^*, y^*) \in H_{x*}$, by assumption (4). Then

$$\min\{\| f(x^*, y^*) - w \| \in \mathbf{R} | w \in H_{x*}\} = 0,$$

and (i) and (iii) become

$$x^* \in \text{co} \bigcup_{\mathcal{T} \in \mathcal{T}} \{\xi \in X | \forall T \in \mathcal{T} : \xi^T \in F^T(x^*) \text{ and } f(x^*, y^*) = u(x^*, \xi)\}$$

$$\subset \text{co} \bigcup_{\mathcal{T} \in \mathcal{T}} \{\xi \in X | \forall T \in \mathcal{T} : \xi^T \in F^T(x^*) \text{ and } f(x^*, y^*) \leq u(x^*, \xi)\}.$$

The last set is identical to

$$\bigcup_{\mathcal{T} \in \mathcal{T}} \{\xi \in X | \forall T \in \mathcal{T} : \xi^T \in F^T(x^*) \text{ and } f(x^*, y^*) \leq u(x^*, \xi)\}$$

by assumption (5), so there exists $\mathcal{T}^* \in \mathcal{T}$ such that

(v) $x^{T*} \in F^T(x^*)$ for every $T \in \mathcal{T}^*$; and

(vi) $f(x^*, y^*) \leq u(x^*, x^*)$.

The coalitional equilibrium condition (1) is satisfied by (v), and so is (2) by (vi) and the definition of $\tau(x^*, y^*)$. □

The balancedness assumption (4) and convexity assumption (5) of Theorem 5.7.1 are probably too obscure to capture since each imposes on the model an implicit relation between F^S and u_S^j. The following theorem shows that an assumption only on F^S and another assumption only on u_S^j jointly imply balancedness and convexity. Denote by $\tilde{F}^S(x)$ the set $\{\xi | \xi^S \in F^S(x), \xi^{N \setminus S} = \mathbf{0}\}$.

5.7. Social Coalitional Equilibrium of a Society

Theorem 5.7.2. *Let* $(\{X^j\}_{j\in N}, \{F^S\}_{S\in \mathcal{N}}, \{u_S^j\}_{j\in S\in \mathcal{N}}, \mathcal{T})$ *be a society. Assume that*

(6) *for any* $x \in X$ *and for any balanced collection* $\mathcal{B}$ *with the associated balancing coefficients* $\{\lambda_S | S \in \mathcal{B}\}$, *it follows that*

$$\sum_{S\in\mathcal{B}} \lambda_S \tilde{F}^S(x) \subset \bigcup_{\mathcal{T}\in\mathcal{T}} \sum_{T\in\mathcal{T}} \tilde{F}^T(x);$$

and

(7) *for each* $j \in N$, *there exists a function* $u^j: X \times X^j \to \mathbf{R}$ *such that* $u^j(x, \cdot)$ *is quasi-concave on* X^j *for any* $x \in X$ *and such that for every* $S \in \mathcal{N}$ *containing* j, $u_S^j(x, \xi^S) = u^j(x, \xi^j)$ *for all* $\xi^S \in F^S(x)$.
Then the society satisfies both balancedness and convexity.

▶PROOF. *Balancedness.* Take any $x \in X$ and any $v \in \mathbf{R}^n$, and suppose there exists a balanced collection $\mathcal{B}$ such that for each $S \in \mathcal{B}$ there exists $\xi(S) \in F^S(x)$ for which $v_j \le u^j(x, \xi^j(S))$ for every $j \in S$. Let $\{\lambda_S\}_{S\in\mathcal{B}}$ be the associated balancing coefficients, and define $\eta \in X$ by

$$\eta^j := \sum_{S\in\mathcal{B}: S \ni j} \lambda_S \xi^j(S) \in X^j.$$

By assumption (7), $v_j \le u^j(x, \eta^j)$ for every $j \in N$, and by assumption (6), there exists $\mathcal{T} \in \mathcal{T}$ such that $\eta^T \in F^T(x)$ for every $T \in \mathcal{T}$.

Convexity. Given $x \in X$ and $v \in \mathbf{R}^n$, take any $\xi, \xi' \in X$ such that $\xi \in \sum_{T\in\mathcal{T}} \tilde{F}^T(x)$, $\xi' \in \sum_{T\in\mathcal{T}'} \tilde{F}^T(x)$ and that $v_j \le u^j(x, \xi^j)$, $v_j \le u^j(x, \xi'^j)$. Take also any $\alpha \in [0, 1]$. Then $\mathcal{B} := \mathcal{T} + \mathcal{T}'$ is a balanced collection, and $\{\lambda_S\}_{S\in\mathcal{B}}$, defined by $\lambda_S = \alpha$ if $S \in \mathcal{T}$, $\lambda_S = 1 - \alpha$ if $S \in \mathcal{T}'$, is a set of associated weights. It is straightforward to check that $\alpha\xi + (1-\alpha)\xi' \in \bigcup_{\mathcal{T}\in\mathcal{T}} \sum_{T\in\mathcal{T}} \tilde{F}^T(x)$ and that $v_j \le u^j(x, \alpha\xi^j + (1-\alpha)\xi'^j)$ for every $j \in N$. □

In order to relate the present discussion to the discussion of Section 4.3, assume that $\mathcal{T}$ consists only of the finest partition $\{\{1\}, \ldots, \{n\}\}$ and that for every $x \in X$ and for every $j \in S \in \mathcal{N}$ with $\#S \ge 2$, $u_{\{j\}}^j(x, \xi^j) > u_S^j(x, \eta^S)$ for all $\xi^j \in F^{\{j\}}(x)$ and all $\eta^S \in F^S(x)$. Then the players do not benefit from cooperation; the present model is reduced to the abstract economy of Section 4.3, and the social coalitional equilibrium is simply the social equilibrium. In this case assumption (4) of Theorem 5.7.1 is automatically satisfied, and assumption (5) is equivalent to [convexity of $F^{\{j\}}(x)$ and quasi-concavity of $u_{\{j\}}^j(x,\cdot)$ on $F^{\{j\}}(x)$ for every $x \in X$ and every $j \in N$]. Thus Theorem 5.7.1 is in this case reduced precisely to Theorem 4.3.1.

Assume, on the other hand, that $\mathcal{T}$ consists only of the coarsest partition $\{N\}$, that for each $S \in \mathcal{N}$, F^S is a constant correspondence (i.e., $F^S(x) = F^S(x')$ for all x, x'), and that for every $j \in S \in \mathcal{N}$, u_S^j is independent of its first argument (i.e., $u_S^j(x,\cdot) = u_S^j(x',\cdot)$ for all x, x') Then what a coalition can do is not influenced by outsiders; the present model is reduced to the non-side-payment game of Section 5.4, and the set of social coalitional equilibria gives rise exactly to the core. Assumptions (1)–(3) of Theorem 5.7.1 correspond roughly to assumptions (1)–(3) of Theorem 5.4.1. Assumption (4) of Theorem 5.7.1 is the balancedness condition, as was noted earlier. Put $\bar{F}^S := F^S(x)$, $\bar{u}_S^j(\xi^S) := u_S^j(x,\xi^S)$. Assumption (5) of Theorem 5.7.1 is the convexity assumption: convexity of $\bar{F}^N$ and quasi-concavity of $\bar{u}_N^j$ on $\bar{F}^N$. Thus in this case Theorem 5.7.1 is weaker than Theorem 5.4.1; the convexity assumption is not used in Theorem 5.4.1.

5.8. Optimality of the Nash Equilibrium: Strong Equilibrium

It has already been observed that a Nash equilibrium of a game in normal form is not necessarily Pareto optimal (Section 4.2). The purpose of this section is to supply conditions under which there exists a Pareto optimal Nash equilibrium.

Let $\{X^j, u^j\}_{j \in N}$ be a game in normal form. To highlight the present problem, use the second interpretation of the game (see the first paragraph of Section 4.1) and define $\tilde{u}^j \colon X \to \mathbf{R}$ by $\tilde{u}^j(x^1, \ldots, x^{j-1}, \xi^j, x^{j+1}, \ldots, x^n) := u^j(x, \xi^j)$. The following convenience will also be adopted by abuse of notation: Given x, $x' \in X$ and $S \in \mathcal{N}$, define $y \in X$ by $y^i := x^i$ if $i \notin S$ and by $y^i := x^{i'}$ if $i \in S$. Then $\tilde{u}^j((x^i)_{i \notin S}, (x^{i'})_{i \in S}) := \tilde{u}^j(y)$. An n-tuple of strategies $x^* \in X$ is called a *strong equilibrium* if it is not true that there exist $S \in \mathcal{N}$ and $(x^i)_{i \in S} \in \prod_{i \in S} X^i$

5.8. Optimality of the Nash Equilibrium: Strong Equilibrium

such that $\tilde{u}^j((x^{i*})_{i\notin S}, (x^i)_{i\in S}) > \tilde{u}^j(x^*)$ for every $j \in S$. In particular, this is a Nash equilibrium that is Pareto optimal.

Theorem 5.8.1. *Let $\{X^j, u^j\}_{j\in N}$ be a game in normal form, and let $\mathcal{T}$ be a nonempty family of coalition structures. There exists a strong equilibrium if*

(1) X^j *is a nonempty, convex, compact subset of a Euclidean space for every $j \in N$;*

(2) $\tilde{u}^j$ *is continuous in X for every $j \in N$;*

(3) *given any $(x, v) \in X \times \mathbf{R}^n$, $\bigl[$there exists a balanced collection $\mathcal{B}$ such that for each $S \in \mathcal{B}$ there exists $(y^{i,S})_{i\in S} \in \prod_{i\in S} X^i$ for which $v_j \leq \tilde{u}^j((x^i)_{i\notin S}, (y^{i,S})_{i\in S})$ for every $j \in S\bigr]$ implies $\bigl[$there exist $\mathcal{T} \in \mathcal{T}$ and $z \in X$ such that $v_j \leq \tilde{u}^j((x^i)_{i\notin T}, (z^i)_{i\in T})$ for all $j \in T \in \mathcal{T}\bigr]$; and*

(4) *given any $(x, v) \in X \times \mathbf{R}^n$, the set $\bigcup_{\mathcal{T}\in\mathcal{T}} \{y \in X \mid v_j \leq \tilde{u}^j((x^i)_{i\notin T}, (y^i)_{i\in T})$ for all $j \in T \in \mathcal{T}\}$ is convex.*

PROOF. This is a straightforward consequence of Theorem 5.7.1. Given the present game in normal form, construct a society $(\{X^j\}_{j\in N}, \{F^S\}_{S\in\mathcal{N}}, \{u_S^j\}_{j\in S\in\mathcal{N}}, \mathcal{T})$ by defining $F^S(x) := \prod_{j\in S} X^j$ for all $x \in X$, $u_S^j(x, \xi^S) := \tilde{u}^j((x^i)_{i\notin S}, (\xi^i)_{i\in S})$ for all $(x, \xi^S) \in X \times X^S$. Then a social coalitional equilibrium of this society is precisely a strong equilibrium. □

The convexity assumption (4) of Theorem 5.8.1 is seldom satisfied when $\#\mathcal{T} \geq 2$. In view of the discussion made between the statement of Theorem 5.7.1 and its proof, one can establish a more useful version:

Corollary 5.8.2. *Let $\{X^j, u^j\}_{j\in N}$ be a game in normal form, and let $\mathcal{T}$ be a coalition structure. There exists a strong equilibrium, if assumptions (1) and (2) of Theorem 5.8.1 are satisfied, if*

(3′) *given any $(x, v) \in X \times \mathbf{R}^n$, $\bigl[$there exists a balanced collection $\mathcal{B}$ such that for each $S \in \mathcal{B}$ there exists $(y^{i,S})_{i\in S} \in \prod_{i\in S} X^i$ for which $v_j \leq \tilde{u}^j((x^i)_{i\notin S}, (y^{i,S})_{i\in S})$ for every $j \in S\bigr]$ implies $\bigl[$there exists $z \in X$ such that $v_j \leq \tilde{u}^j((x^i)_{i\notin T}, (z^i)_{i\in T})$ for all $j \in T \in \mathcal{T}\bigr]$; and if*

(4′) *given any $x \in X$ and any $T \in \mathcal{T}$, the function $\tilde{u}^j((x^i)_{i\notin T}, \cdot)$ is quasi-concave in $\prod_{i\in T} X^i$ for every $j \in T$.*

In order to comment on Corollary 5.8.2, consider the example of the prisoner's dilemma given in Section 4.2. If $\mathscr{T} = \{\{1,2\}\}$, choose $x =$ (first row, first column), $v = (15, 15)$, $\mathscr{B} = \{\{1\}, \{2\}\}$, and $y =$ (second row, second column). Then the hypothesis in assumption (3′) is satisfied, but the conclusion of (3′) is false. Therefore assumption (3′) excludes the prisoner's dilemma. If $\mathscr{T} = \{\{1\}, \{2\}\}$, choose $x =$ (second row, second column), $v = (10, 10)$, $\mathscr{B} = \{\{1, 2\}\}$, and $y =$ (first row, first column). Then the hypothesis in assumption (3′) is satisfied, but the conclusion of (3′) is false. Again assumption (3′) excludes the prisoner's dilemma.

5.9. Notes

Von Neumann and Morgenstern (1947) introduced the side-payment game, but their interpretation of the characteristic function v was specific: Consider a zero-sum n-person game. Using the notation from the second paragraph following the proof of Lemma 4.7.1 in Section 4.7, it is the game that satisfies $\sum_{j \in N} a^j_{h_1 \cdots h_n} = 0$ for every $(h_1, \ldots, h_n) \in \prod_{j \in N} \{1, \ldots, m_j\}$. Given the coalition S, construct a zero-sum two-person game as follows: The two "players" are S and $N \backslash S$, the strategy set for "player S" (for "player $N \backslash S$", resp.) is [the set of probabilities on $\prod_{j \in S} \{1, \ldots, m_j\}$] ([the set of probabilities on $\prod_{j \in N \backslash S} \{1, \ldots, m_j\}$], resp.), and the payoff for "player S" (for "player $N \backslash S$", resp.) is $E \sum_{j \in S} a^j_{h_1 \cdots h_n}$ ($E \sum_{j \in N \backslash S} a^j_{h_1 \cdots h_n}$, resp.), where E is the expectation operator. By the minimax principle, this zero-sum two-person game has the value (see the second paragraph of Section 4.7). The number $v(S)$ was originally interpreted as this value.

The concept of the core of a side-payment game was discovered by Gillies (1959) and Lloyd S. Shapley during their investigation of the von Neumann–Morgenstern solutions in 1952–1953. Bondareva (1962, 1963) and Shapley (1967) independently established Theorem 5.2.1. Denote by $C(v)$ the core of the side-payment game v.

A side-payment game may be identified with a point in the $(2^n - 1)$-dimensional Euclidean space $\mathbf{R}^{\mathcal{N}}$. The *core correspondence* is a corre-

5.9. Notes

spondence $C: \mathbf{R}^{\mathcal{N}} \to \mathbf{R}^n, v \mapsto C(v)$. Since the core is the set of solutions of a linear programming problem (see Exercise 5 of this chapter), a little modification of the argument in Walkup and Wets (1969) establishes.

Theorem 5.9.1. *Let $B(n)$ be the set of all n-person balanced side-payment games identified with a subset of the $(2^n - 1)$-dimensional Eulidean space $\mathbf{R}^{\mathcal{N}}$. Let $C: B(n) \to \mathbf{R}^n$ be the core correspondence. Then there are finitely many continuous piecewise linear functions $c^i: B(n) \to \mathbf{R}^n$, $i \in I$, such that for each $v \in B(n)$, {the extreme points of $C(v)$} $\subset \{c^i(v) | i \in I\} \subset C(v)$.*

In particular, Theorem 5.9.1 shows that C is both u.s.c. and l.s.c. in $B(n)$ since $C(v)$ is compact in $\mathbf{R}^n$.

The theory of the core of a side-payment game with infinitely many players has been well developed. Let A be a set, and let $\mathscr{A}$ be an algebra of subsets of A: The measurable set $(A, \mathscr{A})$ is fixed throughout the present and next two paragraphs. The set A is interpreted as the player set and $\mathscr{A}$ as the family of feasible coalitions. Denote by **ba** the space of all bounded additive scalar functions defined on $\mathscr{A}$. For each $S \in \mathscr{A}$ its *characteristic function* is a function $\chi_S: A \to \mathbf{R}$ defined by $\chi_S(a) = 1$ if $a \in S$ and $\chi_S(a) = 0$ if $a \in A \backslash S$. Given a finite subfamily $\mathscr{B}$ of $\mathscr{A}$ and an indexed set $\{\lambda_S | S \in \mathscr{B}\}$ of nonnegative numbers, one can find the unique finite subfamily $\{T_i\}_{i=1}^m$ of $\mathscr{A}$ and the associated set $\{\gamma_{T_i}\}_{i=1}^m$ of positive numbers such that $T_1 \supsetneq T_2 \supsetneq \cdots \supsetneq T_m$ and that $\sum_{S \in \mathscr{B}} \lambda_S \chi_S = \sum_{i=1}^m \gamma_{T_i} \chi_{T_i}$. The collection $\{\gamma_i, T_i\}_{i=1}^m$ is called the *canonical form of* $\sum_{S \in \mathscr{B}} \lambda_S \chi_S$. A *side-payment game* is a function $v: \mathscr{A} \to \mathbf{R}_+$ such that $v(\emptyset) = 0$. The *core* of game v is the set $\{\mu \in \mathbf{ba} | \mu(A) \leq v(A)$, and $\mu(S) \geq v(S)$ for every $S \in \mathscr{A}\}$ and is denoted by $C(v)$. A game v is called *balanced* if for any finite subfamily $\mathscr{B}$ by $\mathscr{A}$ and any indexed set $\{\lambda_S | S \in \mathscr{B}\}$ of nonnegative numbers for which $\sum_{S \in \mathscr{B}} \lambda_S \chi_S \leq \chi_A$, it follows that $\sum_{S \in \mathscr{B}} \lambda_S v(S) \leq v(A)$. A game v is called *exact* if for any $S \in \mathscr{A}$, there exists $\mu \in C(v)$ such that $\mu(S) = v(S)$. A game v is called *convex* if for any $S, T \in \mathscr{A}$, it follows that $v(S) + v(T) \leq v(S \cup T) + v(S \cap T)$.

Theorem 5.9.2. *Let $v: \mathscr{A} \to \mathbf{R}_+$ be a game. The core of v is nonempty iff game v is balanced.*

Theorem 5.9.3. *Let* $v: \mathscr{A} \to \mathbf{R}_+$ *be a game. Game v is exact iff for any $T \in \mathscr{A}$, it follows that $v(T) = \sup\{\sum_{S \in \mathscr{B}} \lambda_S v(S) - \kappa v(A) | \mathscr{B}$ is a finite subfamily of $\mathscr{A}$, $\{\lambda_S | S \in \mathscr{B}\} \subset \mathbf{R}_+$, $\kappa \in \mathbf{R}_+$, $\sum_{S \in \mathscr{B}} \lambda_S \chi_S - \kappa \chi_A \leq \chi_T\}$.*

Theorem 5.9.4. *Let* $v: \mathscr{A} \to \mathbf{R}_+$ *be a game. The following three conditions are equivalent:*

(1) *Game v is convex;*
(2) *For any finite subfamily $\mathscr{B}^i$ of $\mathscr{A}$ and any indexed set $\{\lambda_S^i | S \in \mathscr{B}^i\}$ of nonnegative numbers, $i = 1, 2$, for which $\sum_{S \in \mathscr{B}^1} \lambda_S^1 \chi_S \leq \sum_{S \in \mathscr{B}^2} \lambda_S^2 \chi_S$, it follows that $\sum_{S \in \mathscr{B}^1} \lambda_S^1 v(S) \leq \sum_{T \in \mathscr{C}} \gamma_T v(T)$, where $\{\gamma_T, T\}_{T \in \mathscr{C}}$ is the canonical form of $\sum_{S \in \mathscr{B}^2} \lambda_S^2 \chi_S$;*
(3) *If $S_1 \supsetneq S_2 \supsetneq \cdots \supsetneq S_m$, where $S_i \in \mathscr{A}$, then there exists $\mu \in C(v)$ such that $\mu(S_i) = v(S_i)$ for all $i = 1, \ldots, m$.*

Schmeidler (1967) established Theorem 5.9.2. Kannai (1969) pointed out that one can prove Theorem 5.9.2 by a direct application of Fan's theorem (Fan, 1956, Theorem 13, pp. 126–128) on a system of linear inequalities. Theorem 5.9.3 is due to Schmeidler (1972); but again, one can prove it by a direct application of Fan's corollary (Fan 1956, Corollary 6, p. 126), as was noted by Delbaen (1974, (2) $\Rightarrow$ (3), p. 216). Rosenmüller (1971) worked on convex games. Delbaen (1974) proved Theorem 5.9.4. Kannai (1969) also provided a necessary and sufficient condition for the core to have a countably additive set function. Schmeidler (1972) identified a necessary and sufficient condition for an exact game to have a core whose members are all countably additive. His technique has proven useful for obtaining the Radon–Nikodym derivative of a member of the core. For a detailed discussion of convex games with finitely many players see Section 6.2.

Denote by **B** (**E**, **C**, resp.) the set of balanced games (the set of exact games, the set of convex games, resp.). The preceding theorems imply: $\mathbf{C} \subset \mathbf{E} \subset \mathbf{B}$. Each inclusion is strict. Endow **B** with the topology induced by the distance $d: d(v, v') := \sup_{S \in \mathscr{A}} |v(S) - v'(S)|$. The space **ba** has the topology induced by the total variation norm. Based on his own result (Theorem 5.9.4), Delbaen (1974) studied the continuity properties of the core correspondence $C: \mathbf{B} \to \mathbf{ba}$. Since the core $C(v)$

5.9. Notes

may not be compact in **ba**, one has to be careful of the definitions of continuity. Recall the definitions of V-upper semicontinuity, V-lower semicontinuity, H-upper semicontinuity, and H-lower semicontinuity (Section 2.1 and Exercise 2 of Chapter 2). Delbaen (1974) proved that *the core correspondence C is H-u.s.c. and V-l.s.c. in* **C**. He also showed by examples that *V-lower semicontinuity of C on* **B** *does not necessarily hold*. Compare these (non)continuity results [for infinite-person games] with the continuity results (Theorem 5.9.1) [for n-person games]. Whether C is u.s.c. and l.s.c. in **E**, and whether C is V-u.s.c. and H-l.s.c. in **C** are still open questions.

Aumann and Peleg (1960) formulated the non-side-payment game, and Aumann (1961) developed the core concept within this framework.

The first successful theorem (Theorem 5.4.1) for nonemptiness of the core was established by Scarf (1967a). He applied the "path-following" technique of Lemke and Howson (1964) and constructed an algorithm for a member of the core. By using the constructive "path-following" technique, one can obtain theorems as deep as or deeper than theorems obtained by using the nonconstructive fixed-point method of Section 3.2; see Scarf (1967b). It is an interesting question whether the fixed-point method (which is very handy!) can generally yield results as deep as those of the "path-following" technique (apart from the computational merit of the latter). The works of Scarf (1967a,b) opened an important field of research in economics: the computation of economic equilibria; see, e.g., Scarf (1973). Also using the "path-following" technique, Shapley (1973) established the K-K-M-S theorem (Theorem 5.3.1), and then applied Theorem 5.3.1 to prove Theorem 5.4.1. Shapley's proof of Theorem 5.4.1 is reproduced in Section 5.4. Ichiishi (1981a) showed that Theorem 5.3.1 can be proved by a simple application of the coincidence theorem (Theorem 3.3.3). The present proof of Theorem 5.3.1 is a variant of the proof by Ichiishi (1981a). Todd (1978) (and his private communication) has given an alternative proof of Theorem 5.3.1 and also a proof of a modified version of Theorem 5.3.1 that uses only an easier nonconstructive fixed-point theorem (Theorem 3.2.2). In this modified version, the assumption of Theorem 5.3.1 is replaced by a stronger condition: For all S, $T \in \mathcal{N}$, $[\Delta^T \cap C_S \neq \emptyset]$ implies $[S \subset T]$. As is evident from the present proof of Theorem 5.4.1, Todd's modified version of Theorem 5.3.1 is enough to prove Theorem 5.4.1. The modified version is not, however, a generalization of the K–K–M theorem (Theorem 3.1.2).

Billera (1970) assumed convexity of the $\tilde{V}(S)$, and using their support functions established a necessary and sufficient condition for non-emptiness of the core of a non-side-payment game V. He also showed an example of a non-side-payment game that is not balanced and yet has a nonempty core.

The theory of the core of a non-side-payment game with infinitely many players is now being developed. Kannai (1969) applied Theorem 5.4.1 to the case in which there are countably many players and provided a sufficient condition for nonemptiness of the core. He did not treat an atomless measure space of players. The following general model was constructed by Ichiishi and Weber (1978): Let $(A, \mathscr{A}, v)$ be a σ-finite, positive measure space of players. Denote by L^∞ the vector lattice of all (equivalence classes modulo v-null sets of) v-essentially bounded $\mathscr{A}$-measurable real-valued functions on A. A member $u \in L^\infty$ is interpreted as a utility allocation as follows: With u understood as a specific function in the equivalence class, player a enjoys his utility level $u(a)$, v-a.e. in A. For each $S \in \mathscr{A}$ define $\text{proj}_S \colon L^\infty \to L^\infty$ by: $(\text{proj}_S u)(a) = u(a)$ if $a \in S$ and $(\text{proj}_S u)(a) = 0$ if $a \in A \setminus S$. Also define

$$L^\infty_+ := \{ u \in L^\infty \,|\, u \geq \mathbf{0} \}.$$

Since the coalitions with positive measure are relevant for the cooperative theory based on a measure space of players, define $\mathscr{A}_0 := \{ S \in \mathscr{A} \,|\, v(S) > 0 \}$. A *non-side-payment game* is a correspondence $V \colon \mathscr{A} \to L^\infty$ satisfying the following:

(1) For all $S, T \in \mathscr{A}$ with $v(T) = 0$, $V(S \setminus T) = V(S)$ and

(2) for all $S \in \mathscr{A}$ and $u, v \in L^\infty$ with $\text{proj}_S u = \text{proj}_S v$, $u \in V(S)$ iff $v \in V(S)$.

The *core* of a non-side-payment game V is the set of all $u^* \in L^\infty$ such that (1) $u^* \in V(A)$ and (2) it is not true that there exist $S \in \mathscr{A}_0$ and $u \in V(S)$ for which $u^*(a) < u(a)$, v-a.e. in S. Ichiishi and Weber (1978) established several necessary and sufficient conditions for nonemptiness of the core by applying Fan's theorem (Fan 1956, Theorem 13, pp. 126–128) repeatedly or else by generalizing this Fan theorem in the context of vector lattices and then applying the somewhat generalized version.

However no sufficient condition that is simple enough to be practical has been found. Ichiishi and Schäffer (1979) introduced a weaker core concept: Given an arbitrary topology $\mathcal{T}$ on L^∞, the $\mathcal{T}$-core is defined as the set $V(A)\setminus\bigcup_{S\in\mathcal{A}_0} \mathring{V}(S)$, where $\mathring{V}(S)$ is the topological interior of $V(S)$ in $(L^\infty, \mathcal{T})$. The following practical theorem (Theorem 5.9.5) is a special case of the main theorem of Ichiishi and Schäffer (1979 revised 1981). For each $u \in L^\infty$, denote by $u_+ \in L^\infty_+$ (*the positive part of u*) the least upper bound in L^∞ of $\{\mathbf{0}, u\}$, i.e., essentially the valuewise maximum of the functions $\mathbf{0}$ and u. If $\mathcal{B}$ is a sub-σ-algebra of $\mathcal{A}$ denote by $L^\infty_\mathcal{B}$ the closed vector sublattice of L^∞ consisting of all $\mathcal{B}$-measurable members of L^∞; this is the closed subspace spanned by $\{\chi_S \mid S \in \mathcal{B}\}$. If $\mathcal{B}$ is a finite subalgebra of $\mathcal{A}$, $L^\infty_\mathcal{B}$ is finite-dimensional. Set $L^\infty_{\mathcal{B}+} := L^\infty_+ \cap L^\infty_\mathcal{B}$. A finite subcollection $\mathcal{B}$ of $\mathcal{A}$ is called *balanced* if $\chi_A = \sum_{S\in\mathcal{B}} \lambda_S \chi_S$ for a suitable family $\{\lambda_S \mid S \in \mathcal{B}\}$ in $\mathbf{R}_+$. A game V is called *balanced* if $\bigcap_{S\in\mathcal{B}} V(S) \subset V(A)$ for every balanced finite subcollection $\mathcal{B}$ of $\mathcal{A}$.

Theorem 5.9.5. *Given a probability measure space $(A, \mathcal{A}, \nu)$ of players, let $V: \mathcal{A} \to L^\infty$ be a non-side-payment game such that $V(S) \cap L^\infty_+ \neq \emptyset$ for every $S \in \mathcal{A}$ and such that $V(A)$ is $\mathcal{T}$-closed. Let $\mathcal{T}$ be a Hausdorff vector-space topology that is either stronger than the norm-topology or weaker than the norm-topology, but strong enough so that L^∞_+ is $\mathcal{T}$-closed. The $\mathcal{T}$-core is nonempty if*

(1) $V(S) - L^\infty_+ = V(S)$ *for every $S \in \mathcal{A}$;*

(2) *for every finite subalgebra $\mathcal{F}$ of $\mathcal{A}$ there is a number $\mu_\mathcal{F} \in \mathbf{R}_+$ such that for every $S \in \mathcal{F}$ every $u \in V(S) \cap L^\infty_{\mathcal{F}+}$ satisfies $[u(a) < \mu_\mathcal{F}$, ν-a.e. in $S]$;*

(3) *there exists a $\mathcal{T}$-compact subset K of $V(A) \cap L^\infty_+$ such that, for every finite subalgebra $\mathcal{F}$ of $\mathcal{A}$,*

$$L^\infty_{\mathcal{F}+} \cap V(A) \setminus \bigcup_{S\in\mathcal{F}\cap\mathcal{A}_0} \mathring{V}(S) \subset (K - L^\infty_+) \cap L^\infty_+; \text{ and}$$

(4) *game V is balanced.*

Assumption (3) may be replaced by a stronger one that is easier to verify:

(3′) There exists a $\mathcal{T}$-compact subset K of L_+^∞ such that

$$V(A) \cap L_+^\infty = (K - L_+^\infty) \cap L_+^\infty.$$

When $\mathcal{T}$ is chosen to be the weak* topology $\sigma(L^\infty, L^1)$, however, replacement of assumption (3) by (3′) makes the theorem trivial because assumptions (3′) and (4) together imply that there exists a number $\mu \in \mathbf{R}$ such that for every $S \in \mathcal{A}_0$, every $u \in V(S) \cap L_+^\infty$ satisfies $[u(a) < \mu, \text{ } v\text{-a.e. in } S]$ and consequently that $L_+^\infty \cap \overset{\circ}{V}(S) = \emptyset$ for every $S \in \mathcal{A}_0$ if v is atomless. The present theorem is nontrivial even when $\mathcal{T}$ is the weak* topology; an example is given in Ichiishi and Schäffer (1979 revised 1981). When player set A is finite, so that L^∞ is finite-dimensional, Theorem 5.9.5 reduces precisely to Theorem 5.4.1. Given any positive number $\epsilon > 0$, the ϵ-core of a non-side-payment game V is the set of all $u^* \in L^\infty$ such that (1) $u^* \in V(A)$; and that (2) it is not true that there exist $S \in \mathcal{A}_0$ and $u \in V(S)$ for which $u^*(a) + \epsilon < u(a)$, v-a.e. in S. Under assumption (1) of Theorem 5.9.5, when $\mathcal{T}$ is chosen to be the norm-topology, the $\mathcal{T}$-core is identical to $\bigcap_{\epsilon > 0}$ (the ϵ-core). S. Weber (1981) studied independently the set $\bigcap_{\epsilon > 0}$ (the ϵ-core), which he called the *weak core*. He supplied a sufficient condition for nonemptiness of the weak core. The S. Weber theorem is somewhat stronger than the special case of Theorem 5.9.5 in which $\mathcal{T}$ is chosen to be the norm-topology, in the sense that he assumes only σ-finiteness of the measure v (rather than finiteness of v as in Theorem 5.9.5). However most work on economic or game-theoretical models that uses measure theory, such as Hildenbrand (1974) and Aumann and Shapley (1974), assumes finiteness of the measure v. The S. Weber theorem is weaker than the special case of Theorem 5.9.5 in which $\mathcal{T}$ is chosen to be the norm-topology, in the sense that he assumes "strong balancedness" rather than balancedness. Strong balancedness restricts economic applicability of the model; see Exercise 6 of this Chapter. Rosenmüller (1975) considered a model in which A is a Polish space and $\mathcal{A}$ is the σ-algebra of Borel subsets of A. In his setup, the space of utility allocations is not L^∞ but rather the space of all Radon measures on A endowed with the weak* topology. His core concept is very close, in spirit, to the preceding $\mathcal{T}$-core concept when $\mathcal{T}$ is the weak* topology.

Lemma 5.5.1 and Theorem 5.5.2 are due to Scarf (1967a).

The concept of core allocation within the framework of a pure exchange economy goes as far back as Edgeworth (1881). There, using a model of pure exchange with two commodities and two types of consumers, he asserted that a competitive allocation is a core allocation and that the set of core allocations "converges" to the set of competitive allocations as the number of consumers of each type goes to infinity. The Edgeworth proposition has drawn the attention of economists during the past two decades due to the pioneering work of Shubik (1959). The intuitive exposition in Section 5.6 that makes use of Figure 5.6.1, is taken from Debreu and Scarf (1972). Two successful theorems established in the early stage of the investigations of the Edgeworth proposition are as follows: A limit theorem by Debreu and Scarf (1963) on replica economies and an equivalence theorem by Aumann (1964) on an economy with an atomless measure space of consumers. The former is presented here first.

Let $\mathscr{E} := \{\mathbf{R}_+^l, \lesssim_i, \omega^i\}_{i=1}^m$ be a pure exchange economy. The q-replica of the economy $\mathscr{E}$ is the pure exchange economy $\mathscr{E}^{(q)}$ with qm consumers in which there are q agents having consumer characteristics $(\mathbf{R}_+^l, \lesssim_i, \omega^i)$ for every $i = 1, \ldots, m$. Debreu and Scarf (1963) first established that under mild assumptions on $\mathscr{E}$, any core allocation $(x^{i,k*})_{i=1, k=1}^{m, q}$ of $\mathscr{E}^{(q)}$ has the *equal treatment property*: For every i, $x^{i,k*} = x^{i*}$ for all $k = 1, \ldots, q$ (Exercise 7). Then the core allocation $(x^{i,k*})_{i,k}$ of $\mathscr{E}^{(q)}$ may be represented by the m-tuple of members of $\mathbf{R}_+^l$, $(x^{i*})_{i=1}^m$. Denote by $C(\mathscr{E}^{(q)})$ the set of all core allocations of $\mathscr{E}^{(q)}$ represented by points of $\mathbf{R}_+^{lm}$

$$\{(x^{i*})_{i=1}^m \in \mathbf{R}_+^{lm} \mid \underbrace{(x^{i*}, \ldots, x^{i*})}_{q} {}_{i=1}^m \text{ is a core allocation of } \mathscr{E}^{(q)}\}.$$

Clearly,

$$C(\mathscr{E}^{(q)}) \supset C(\mathscr{E}^{(q+1)}) \quad \text{for every } q.$$

Theorem 5.9.6. *Let* $\mathscr{E} := \{\mathbf{R}_+^l, \lesssim_i, \omega^i\}_{i=1}^m$ *be a pure exchange economy. Assume that* $\lesssim_i$ *is complete, transitive, closed, monotone, and strictly convex, and assume that* $\omega^i \gg \mathbf{0}$. *Then for any* $(x^{i*})_{i=1}^m \in \bigcap_q C(\mathscr{E}^{(q)})$ *there exists* $p^* \in \Delta^L$ *such that* $((x^{i*})_{i=1}^m, p^*)$ *is a competitive equilibrium of* $\mathscr{E}$ *(hence of* $\mathscr{E}^{(q)}$ *for any* q*).*

To prove the preceding Debreu–Scarf theorem, first check that all of the assumptions for the equal treatment property (Exercise 7) are satisfied. Apply Anderson's corollary (Corollary 5.6.3) to the sequence of economies defined by $\mathscr{E}^q := \mathscr{E}^{(q)}$; there exists a sequence of price vectors $\{p^{q*}\}_q$ in Δ^L with properties (i) and (ii) of Corollary 5.6.3. Let p^* be an accumulation point of $\{p^{q*}\}_q$. Then for every i,

(i') $p^* \cdot x^{i*} = p^* \cdot \omega^i$ and
(ii') $\inf\{p^* \cdot x \mid x >_i x^{i*}\} = p^* \cdot \omega^i$.

The required result follows directly from (i') and (ii').

The other early successful result on the Edgeworth proposition, the equivalence theorem by Aumann (1964), is concerned directly with a "limit economy" having a probability measure space of consumers $(A, \mathscr{A}, v)$. It is an idealized economic model, in which consumer set A typically has the cardinality of the continuum (i.e., the economy consists of "many" agents), and the probability measure v is atomless (i.e., the relative weight of each agent in the total population is "small").

Theorem 5.9.7. *Let* $\mathscr{E} := \{\mathbf{R}_+^l, \lesssim_a, \omega(a)\}_{a \in A}$ *be a pure exchange economy with an atomless probability measure space of consumers* $(A, \mathscr{A}, v)$. *Assume that the preference relations* $\lesssim_a$ *are monotone, closed, and measurable* $[$*for any* $f, g \in (L_+^1)^l$ *the set* $\{a \in A \mid f(a) >_a g(a)\}$ *is a member of* $\mathscr{A}]$, *and assume that* $\omega \in (L_+^1)^l$ *and* $\int_A \omega \, dv \gg \mathbf{0}$. *Then the competitive allocations of* $\mathscr{E}$ *are precisely the core allocations of* $\mathscr{E}$.

Further work was done on the Edgeworth proposition, in particular by Vind (1965), Kannai (1970), Jean-François Mertens, Hildenbrand (1970), and Bewley (1973); each contribution plays a key role in a treatise by Hildenbrand (1974). Two approaches by these authors should be distinguished. One approach, originated by Vind (1965), studies *one* economy $\mathscr{E}$ with finitely many agents and obtains an upper bound for "noncompetitiveness" (suitably defined) of core allocations of $\mathscr{E}$. The other approach, originated in Kannai (1970) (the first version of which had been circulated since 1964) and followed by all of the authors listed above (except Vind), studies a *sequence* of economies $\{\mathscr{E}^q\}_q$ with finitely many agents that "converges" to its "limit economy." A typical study object here is the so-called *purely competitive sequence of economies* and its "limit economy"; see, e.g., Hildenbrand (1974, p. 138) for a definition. The most successful theorem of the latter approach is due to Bewley (1973), which asserts that *given a purely competitive sequence $\{\mathscr{E}^q\}_q$ with strictly convex preference relations, for any $\epsilon > 0$ there exists $\overline{q}$ such that for each $q \geq \overline{q}$ and every core allocation $(x^{i*})_i$ of $\mathscr{E}^q$, there exists a competitive equilibrium price vector p^* of the "limit economy" with the property that $\|x^{i*} - \xi^i(p^*)\| < \epsilon$ for all agents i of $\mathscr{E}^q$, where $\xi^i(p^*)$ is the demand by agent i under the price vector p^*.*

Arrow and Hahn (1971) studied an economy without a convexity assumption on the preference relations. Applying a version of the Shapley–Folkman theorem (Theorem 1.6.5), they established a result that now serves as the primitive form of Anderson's theorem (Theorem 5.6.2). Hildenbrand (1974) extended the purely competitive sequence approach to economies without convex preference relations. His result may be considered the primitive form of Anderson's corollary (Corollary 5.6.3); because of his explicit use of the "limit economy," however, Hildenbrand can choose competitive equilibrium price vectors of the "limit economy" for the p^{q*} of Corollary 5.6.3. Anderson (1981) also applied his theorem (Theorem 5.6.2) to the situations in which the preference relations are strictly convex.

Section 5.7 is based on Ichiishi (1981b). For an application of Theorem 5.7.1 to the theory of the firm, as suggested in the second paragraph of Section 5.7, see Ichiishi (1982b, 1979). Ichiishi and Quinzii (1981) also use the same logic in their analysis of a production economy with increasing returns.

The concept of strong equilibrium was first introduced by Aumann (1959). The strong equilibrium existence theorems (Theorem 5.8.1 and Corollary 5.8.2) are established in Ichiishi (1982a). Closely related to the strong equilibria are the α-core and β-core of a game in normal form. Both terms are due to Aumann (1961). A member of the β-core was once called an acceptable payoff vector by Aumann (1959). Let $\{X^j, \tilde{u}^j\}_{j \in N}$ be an n-person game in normal form as defined in Section 5.8. Define $X^S := \prod_{j \in S} X^j$ for each $S \in \mathcal{N}$, and set $X := X^N$. Choose a feasible utility allocation $v^* \in \mathbf{R}^N$ (i.e., there exists $x^* \in X$ such that $v_j^* \leq \tilde{u}^j(x^*)$ for every $j \in N$). The utility allocation v^* is in the α-*core* if it is not true that there exist $S \in \mathcal{N}$ and $x^S \in X^S$ such that for any $y^{N \setminus S} \in X^{N \setminus S}$, $\tilde{u}^j(x^S, y^{N \setminus S}) > v_j^*$ for every $j \in S$. In other words, an α-core utility allocation is feasible via the grand coalition and is such that no coalition has its joint strategy which guarantees higher utility levels for all of its members independent of the actions of the outsiders. The utility allocation v^* is in the β-core if it is not true that there exists $S \in \mathcal{N}$ such that for any $y^{N\setminus S} \in X^{N\setminus S}$ there exists $x^S \in X^S$ for which $\tilde{u}^j(x^S, y^{N\setminus S}) > v_j^*$ for every $j \in S$. A β-core utility allocation is feasible via the grand coalition and is such that any coalition can be prevented by its complementary coalition from obtaining higher utility levels for all of its members. From the definitions, it is straightforward to check the following: *The set of strong equilibrium utility allocations is a subset of the β-core, and the β-core is a subset of the α-core.* Thus the assumptions of the strong equilibrium existence theorem (Corollary 5.8.2) also guarantee nonemptiness of the β-core and of the α-core. The larger a given set, the more likely that it is nonempty. One expects, therefore, that a weaker set of assumptions guarantees nonemptiness of the α-core. Indeed Scarf (1971) has established the following theorem on the α-core by applying his earlier theorem (Theorem 5.4.1):

Theorem 5.9.8. *Let $\{X^j, \tilde{u}^j\}_{j \in N}$ be an n-person game in normal form. The α-core is nonempty if for every $j \in N$, X^j is a nonempty, convex, compact subset of a Euclidean space and $\tilde{u}^j$ is continuous and quasi-concave in X.*

Scarf also constructed a simple example in which the β-core is empty.

Some other work related to the core of a game in characteristic function form is noted: (1) Aubin (1979) studied the core of a game, side-payment or non-side-payment, with fuzzy coalitions. (2) In the first paragraph of Section 5.5, a non-side-payment game was constructed from a given pure exchange economy. Billera (1974), and also some of his work referenced therein that was done jointly with R. E. Bixby, asked the converse: Given a non-side-payment game, can one construct a pure exchange economy whose associated non-side-payment game is precisely the given one? (3) Kalai and Schmeidler (1977) studied a broadly applicable concept, the *admissible set*. Given a non-side-payment game satisfying a mild condition, the admissible set is identical to the core. Given a game in normal form, the admissible set contains the Nash equilibria.

EXERCISES

1. Consider a system of linear inequalities with a nonnegativity constraint:
$$Ax \leq b, \quad x \geq 0. \quad (1)$$
Prove that (1) has a solution iff for every $p \in \mathbf{R}^m$ such that $p^t A \geq 0$ and that $p \geq 0$, it is true that $p^t b \geq 0$.

2. Consider a system of linear inequalities with nonnegativity constraint:
$$p^t A \geq c^t, \quad p \geq 0. \quad (2)$$
Prove that (2) has a solution $p \in \mathbf{R}^m$ iff for every $x \in \mathbf{R}^n$ such that $Ax \leq 0$ and that $x \geq 0$, it is true that $c^t x \leq 0$.

3. Given an $m \times n$ matrix A and vectors $b \in \mathbf{R}^m$ and $c \in \mathbf{R}^n$, consider the following two problems*: a primary linear programming problem (P) and its dual linear programming problem (D):

$$\begin{aligned} \text{maximize:} \quad & c^t y \\ \text{subject to:} \quad & Ay \leq b, \quad y \geq 0. \end{aligned} \quad \text{(P)}$$

$$\begin{aligned} \text{minimize:} \quad & x^t b, \\ \text{subject to:} \quad & x^t A \geq c^t, \quad x \geq 0. \end{aligned} \quad \text{(D)}$$

(i) Prove that for every $y \in \mathbf{R}^n$ such that $Ay \leq b$ and $y \geq 0$, and for every $x \in \mathbf{R}^m$ such that $x^t A \geq c^t$ and $x \geq 0$, it is true that $x^t b \geq c^t y$.

(ii) Consider a system of linear inequalities with a nonnegativity constraint
$$\begin{bmatrix} 0 & A \\ -A^t & 0 \\ b^t & -c^t \end{bmatrix} \begin{bmatrix} x \\ y \end{bmatrix} \leq \begin{bmatrix} b \\ -c \\ 0 \end{bmatrix}, \quad \begin{bmatrix} x \\ y \end{bmatrix} \geq 0. \quad (1)$$

Prove that if $(\bar{x}, \bar{y})$ is a solution of (1), then $\bar{x}^t b = c^t \bar{y}$.

* The pair of statements (i) and (vii) is called the *duality theorem* of linear programming.

(iii) Using the same notation as that in (ii), prove that $\bar{y}$ is an optimal solution of (P) and that $\bar{x}$ is an optimal solution of (D).

(iv) Throughout (iv)–(vi) let $(p, q, \alpha) \in \mathbf{R}^m \times \mathbf{R}^n \times \mathbf{R}$ be any vectors such that

$$[p^t, q^t, \alpha] \begin{bmatrix} 0 & A \\ -A^t & 0 \\ b^t & -c^t \end{bmatrix} \geq \mathbf{0}, \qquad [p^t, q^t, \alpha] \geq \mathbf{0}.$$

Prove that if $\alpha \neq 0$, then

$$[p^t, q^t, \alpha] \begin{bmatrix} b \\ -c \\ 0 \end{bmatrix} \geq 0.$$

(*Hint:* Apply (i) to vectors p/α and q/α.)

(v) Prove that if $\alpha = 0$, then

$$p^t A \geq \mathbf{0}, \qquad p \geq \mathbf{0}, \qquad Aq \leq \mathbf{0}, \qquad q \geq \mathbf{0}.$$

(vi) Assume that there exist $y \in \mathbf{R}^n$ such that $Ay \leq b$ and $y \geq \mathbf{0}$ and $x \in \mathbf{R}^m$ such that $x^t A \geq c^t$ and $x \geq \mathbf{0}$. If $\alpha = 0$, then

$$[p^t, q^t, \alpha] \begin{bmatrix} b \\ -c \\ 0 \end{bmatrix} \geq 0.$$

(vii) Prove that if there exists $y \in \mathbf{R}^n$ that satisfies the constraints of (P) and if there exists $x \in \mathbf{R}^m$ that satisfies the constraints of (D), then both (P) and (D) have optimal solutions. For any optimal solution $\bar{y}$ of (P) and for any optimal solution $\bar{x}$ of (D), it is true that $\bar{x}^t b = c^t \bar{y}$.

(viii) Honor. Prove that if there exists $y \in \mathbf{R}^n$ that satisfies the constraint of (P) and if there exists $M \in \mathbf{R}$ such that $c^t y \leq M$ for any such y, then there exists $x \in \mathbf{R}^m$ that satisfies the constraint of (D). Use this fact to prove that (P) has an optimal solution iff (D) has an optimal solution.

4. Using the same notation as that in Exercise 3, consider a primary linear programming problem (P′) with "equality" constraints and the dual problem (D′) without a nonnegativity constraint:

$$\text{maximize:} \quad c^t y$$
$$\text{subject to:} \quad Ay = b, \quad y \geq \mathbf{0}. \tag{P′}$$

$$\text{minimize:} \quad x^t b$$
$$\text{subject to:} \quad x^t A \geq c^t. \tag{D′}$$

Prove that the duality theorem holds for (P′) and (D′), namely,

(i) for every feasible solution $y \in \mathbf{R}^n$ of (P′) and for every feasible solution $x \in \mathbf{R}^m$ of (D′) it follows that $x^t b \geq c^t y$;

(ii) if both (P′) and (D′) are consistent, then both (P′) and (D′) have optimal solutions; and

(iii) for any optimal solution $\bar{y}$ of (P′) and for any optimal solution $\bar{x}$ of (D′) it follows that $\bar{x}^t b = c^t \bar{y}$.

(*Hint:* The constraints of (P′) are

$$\begin{bmatrix} A \\ -A \end{bmatrix} y \leq \begin{bmatrix} b \\ -b \end{bmatrix}, \quad y \geq \mathbf{0}.)$$

5. Let $v: \mathcal{N} \to \mathbf{R}$ be an n-person side-payment game. Recall that χ_S is is the characteristic vector of S for each $S \subset N$. Denote by A the $n \times (2^n - 1)$ matrix whose S-column is χ_S for $S \in 2^N \setminus \{\emptyset\}$. Let c be the vector in $\mathbf{R}^{2^n-1}$ whose S-component is $v(S)$, and let $b := \chi_N$. Given these specific matrices A, c, and b, consider problems (P′) and (D′) of Exercise 4.

(i) Prove that the core of v is nonempty iff problem (D′) has the optimal value equal to $v(N)$.

(ii) Using the duality theorem, give an alternative proof of the Bondareva–Shapley theorem (Theorem 5.2.1).

6. Let $V: \mathcal{N} \to \mathbf{R}^n$ be a non-side-payment game. Recall that $\tilde{V}(S) := V(S) \cap \mathbf{R}^S$. Game V is called *strongly balanced* if for every balanced subfamily $\mathcal{B}$ of $\mathcal{N}$ with the associated balancing coefficients $\{\lambda_S | S \in \mathcal{B}\}$, it follows that $\sum_{S \in \mathcal{B}} \lambda_S \tilde{V}(S) \subset \tilde{V}(N)$.

(i) Show that a strongly balanced game is balanced.
(ii) Let $\{X^j, \lesssim_j, \omega^j\}_{j \in N}$ be a pure exchange economy in which each preference relation $\lesssim_j$ is represented by a utility function $u^j \colon X^j \to \mathbf{R}$. Prove that if X^j is a convex subset of $\mathbf{R}^l$ and u^j is a *concave* function, for every $j \in N$, then the associated game V is strongly balanced.

7. Let $\mathscr{E} := \{\mathbf{R}_+^l, \lesssim_i, \omega^i\}_{i=1}^m$ be a pure exchange economy. Assume that $\lesssim_i$ is complete, transitive, closed, monotone, and strictly convex. Let $\mathscr{E}^{(q)}$ be the q-replica of the economy $\mathscr{E}$, and let $(x^{i,k*})_{i=1,k=1}^{m,\ q}$ be a core allocation of $\mathscr{E}^{(q)}$.

   (i) For each type i, denote by x^i the least desired member of $(x^{i,k*})_k$ with respect to $\lesssim_i$, and set $\bar{x}^i := \sum_{k=1}^q x^{i,k*}/q$. Show $x^i \lesssim_i \bar{x}^i$ and $\sum_{i=1}^m \bar{x}^i = \sum_{i=1}^m \omega^i$.
   (ii) Let S be a coalition of m members consisting of one consumer of each type, the one of type i receiving x^i. Show that if there exists i_0 such that $x^{i_0} \neq x^{i_0,k*}$ for some k, then $\bar{x}^{i_0} >_{i_0} x^{i_0}$.
   (iii) Prove the equal treatment property of the core allocation $(x^{i,k*})_{i,k}$: For every i, $x^{i,k*} = x^{i*}$ for all $k = 1, \ldots, q$.

6

Cooperative Behavior and Fairness

The central concepts of this chapter are the Shapley value of a side-payment game and the λ-transfer value of a non-side-payment game. These are cooperative solution concepts that characterize fairness. Shapley's seminal theorem on the existence and uniqueness of the Shapley value is presented in Section 6.1 (Theorem 6.1.1). For a class of side-payment games (more precisely, for convex games), the Shapley value and the core (see Section 5.2) are closely related. The relationship is explicitly presented in Section 6.2. Several authors have generalized the Shapley value by proposing new solution concepts for non-side-payment games. In Section 6.3 one of the generalized concepts, due to Shapley, is presented and its existence theorem is proved (Theorem 6.3.1); the generalized value is customarily called a λ-transfer value. A value allocation is a commodity allocation in an economy which gives rise to a λ-transfer value. A value allocation existence theorem (Theorem 6.4.1) is proved in Section 6.4. Champsaur's contribution to the Aumann–Shapley proposition is presented in Section 6.5; the Aumann–Shapley proposition establishes a close relationship between competitive allocations and value allocations in a "large" pure exchange economy.

6.1. Shapley Value of a Side-Payment Game

Given the set of players N, and hence given the set of nonempty coalitions of players $\mathcal{N}$, a side-payment game is defined as a function $v: \mathcal{N} \to \mathbf{R}$. Another important cooperative solution concept is now given; it characterizes fairness of the outcome. Denote by G^N the set of all side-payment games with player set N. The set G^N is the set of all real-valued functions defined on $\mathcal{N}$. For every $v, w \in G^N$, $v + w$ is a game defined by $(v + w)(S) := v(S) + w(S)$ for all $S \in \mathcal{N}$. For every $v \in G^N$ and every $r \in \mathbf{R}$, rv is also a game defined by $(rv)(S) := rv(S)$. Thus G^N becomes a vector space over $\mathbf{R}$. For every $v \in G^N$ and every permutation σ on N, $\sigma_* v$ is a game defined by $(\sigma_* v)(S) := v(\sigma(S))$; coalition S plays the same role in game $\sigma_* v$ as coalition $\sigma(S)$ does in game v. A *null player* of a game v is a member j of N such that he does neither good nor bad to any coalition; more precisely, $v(\{j\}) = 0$, and $v(S \cup \{j\}) = v(S)$ for every $S \in \mathcal{N}$.

A *value* on G^N is a function φ from G^N to $\mathbf{R}^n$ such that (1) φ is linear; (2) $(\varphi(\sigma_* v))_j = (\varphi v)_{\sigma(j)}$ for every permutation σ on N, every $v \in G^N$ and every $j \in N$; (3) $\sum_{j \in N} (\varphi v)_j = v(N)$; and (4) $(\varphi v)_j = 0$ if j is a null player of v. The vector φv is interpreted as an outcome of game v; if the players N play game v, then the jth player receives $(\varphi v)_j$ as his payoff. Condition (3) means that this outcome is feasible and (Pareto) efficient. Condition (4) is called the dummy axiom. The meaning of the linearity condition (1) is straightforward. It is the symmetry condition (2) that makes the outcome function φ fair. Player j plays precisely the same role in game $\sigma_* v$ as player $\sigma(j)$ does in game v, so the payoff that player j receives by playing game $\sigma_* v$ should be the same as the payoff that player $\sigma(j)$ receives by playing game v. For each game $v \in G^N$ define $v(\emptyset) := 0$. Denote by G_n the symmetric group on N; i.e., the set of all permutations on N.

Theorem 6.1.1. *There exists one and only one value on G^N. The value is explicitly written as*

$$(\varphi v)_j = \sum_{S \subset N \setminus \{j\}} \gamma_S [v(S \cup \{j\}) - v(S)] \tag{\dag}$$

6.1. Shapley Value of a Side-Payment Game

where

$$\gamma_S := (\#S)! \cdot (n - \#S - 1)!/(n!)$$

or, equivalently,

$$(\varphi v)_j = \sum_{\sigma \in G_n} \frac{1}{n!} [v(P_j^\sigma \cup \{j\}) - v(P_j^\sigma)], \qquad (\dagger\dagger)$$

where $P_j^\sigma := \{i \in N \mid \sigma(i) < \sigma(j)\}$ is the set of all players who precede j with respect to the order σ.

PROOF. Equivalence of formulas ($\dagger$) and ($\dagger\dagger$) is straightforward.

Existence. It suffices to check that the function $\varphi: G^N \to \mathbf{R}^n$, defined as in ($\dagger\dagger$), satisfies all four conditions of the value, which is immediate.

Uniqueness. Step 1. For each $T \in \mathcal{N}$ define a game $v_T \in G^N$ by

$$v_T(S) := \begin{cases} 1 & \text{if } S \supset T, \\ 0 & \text{otherwise.} \end{cases}$$

Then for any value φ,

$$(\varphi v_T)_j = \begin{cases} 1/\#T & \text{if } j \in T, \\ 0 & \text{otherwise.} \end{cases}$$

Indeed, any $j \notin T$ is a null player, so by the dummy axiom (4), $(\varphi v_T)_j = 0$ for such j. For any $i, j \in T$, consider a permutation σ on N defined by $\sigma(k) = k$ if $k \neq i, j$; $\sigma(i) = j$; and $\sigma(j) = i$. Then by the symmetry condition (2), $(\varphi v_T)_i = (\varphi(\sigma_* v_T))_i = (\varphi v_T)_{\sigma(i)} = (\varphi v_T)_j$. The required result of this step now follows from the Pareto optimality condition (3): $\sum_{j \in N} (\varphi v_T)_j = v_T(N) = 1$.

Step 2. Clearly G^N is $(2^n - 1)$-dimensional. It will be shown that a basis for G^N is $\{v_T \mid T \in \mathcal{N}\}$. It suffices to prove its linear independence. Suppose that the set $\{v_T \mid T \in \mathcal{N}\}$ is linearly dependent; then there exist real numbers $\{r_T \mid T \in \mathcal{N}\}$, not all zero, such that $\sum_{T \in \mathcal{N}} r_T v_T$ maps every S to 0. Let $\mathcal{N}' := \{T \in \mathcal{N} \mid r_T \neq 0\}$, and choose $T_0 \in \mathcal{N}'$ such that $\#T_0 \leq \#T$ for all $T \in \mathcal{N}'$. Define $s_T := -r_T/r_{T_0}$, for all $T \in \mathcal{N}'' := \mathcal{N}' \setminus \{T_0\}$. Then $v_{T_0} = \sum_{T \in \mathcal{N}''} s_T v_T$. Now $v_{T_0}(T_0) = 1$. On the other hand, $v_T(T_0) = 0$ for all $T \in \mathcal{N}''$—a contradiction.

Step 3. It was shown in Step 1 that the outcome φv_T does not depend on the specific choice of a value φ. Since $\{v_T \mid T \in \mathcal{N}\}$ is a basis for G^N by Step 2, linearity of a value φ establishes that for any $v \in G^N$ the outcome φv does not depend on the specific choice of φ. □

The unique value is customarily called the *Shapley value*; it will be denoted by φ throughout the rest of this chapter. Given a game $v \in G^N$, φv is called the *Shapley value of v*. The term $v(P_j^\sigma \cup \{j\}) - v(P_j^\sigma)$ is the marginal worth of player j to the coalition P_j^σ. Recall that $\#G_n = n!$. If each order (permutation) is to form with probability $1/(n!)$, then the expectation of marginal worth is precisely the Shapley value. To define the Shapley value as the fair outcome of a game, one needs the concept of vector space of *all* games G^N. To define the Shapley value as the expected marginal worth, one applies formula (††) to *a* given game.

A side-payment game v is called *monotone* if $[S \subset T]$ implies $[0 \le v(S) \le v(T)]$. From formula (†), it is straightforward that

Corollary 6.1.2. *For a monotone game* $v \in G^N$, $(\varphi v)_j \ge 0$ *for all* $j \in N$.

6.2. Convex Game

One cannot hope for a general relationship between the core and the Shapley value. A nonbalanced game, for example, has an empty core and yet has the Shapley value associated with it according to formula (††). There is a certain class of games, however, within which the two cooperative solution concepts are closely related; this class is the subject of this section.

Given a side-payment game $v: \mathcal{N} \to \mathbf{R}$ and a permutation $\sigma \in G_n$, define a vector $a^\sigma(v) \in \mathbf{R}^n$ by

$$a_j^\sigma(v) := v(P_j^\sigma \cup \{j\}) - v(P_j^\sigma), \quad \text{for every} \quad j \in N,$$

where $P_j^\sigma := \{i \in N \mid \sigma(i) < \sigma(j)\}$ is the set of all players who precede j with respect to the order σ. The vector $a^\sigma(v)$ is called a *marginal worth vector*. Then

$$\varphi v = \sum_{\sigma \in G_n} \frac{1}{n!} \cdot a^\sigma(v); \qquad (††)$$

6.2. Convex Game

in particular, the Shapley value is the average of the marginal worth vectors.

A side-payment game $v: \mathcal{N} \to \mathbf{R}$ is called *convex* if for every $S, T \in \mathcal{N}$

$$v(T \cup S) + v(T \cap S) \geq v(T) + v(S).$$

[Recall the convention that $v(\emptyset) = 0$.] The following discussion justifies this terminology: Fix any $R \in \mathcal{N}$, and define the "difference operator" $\Delta_R : G^N \to G^N$ by

$$[\Delta_R v](S) := v(S \cup R) - v(S \backslash R).$$

Given $Q, R \in \mathcal{N}$, the "second-order difference operator" $\Delta_{QR} : G^N \to G^N$ is defined as $\Delta_{QR} v := \Delta_Q(\Delta_R v)$. Now it is easy to check that game v is convex iff for all $Q, R, S \in \mathcal{N}$ it follows that $[\Delta_{QR} v](S) \geq 0$. This last condition is analogous to the standard condition of convexity of a function from $\mathbf{R}$ to $\mathbf{R}$.

A side-payment game $v: \mathcal{N} \to \mathbf{R}$ is said to satisfy *increasing returns with respect to the coalition size* if for any $A, B \subset N$ and any $j \in N$ such that $A \subset B \subset N \backslash \{j\}$, it follows that

$$v(B \cup \{j\}) - v(B) \geq v(A \cup \{j\}) - v(A).$$

Theorem 6.2.1. *Let $v: \mathcal{N} \to \mathbf{R}$ be a game. Then the following four conditions are equivalent:*

(i) *The game v is convex.*
(ii) *For any $A, B, C \subset N$ such that $A \subset B \subset N \backslash C$,*

$$v(B \cup C) - v(B) \geq v(A \cup C) - v(A).$$

(iii) *The game v satisfies increasing returns with respect to the coalition size.*
(iv) *For any $\sigma \in G_n$, the marginal worth vector $a^\sigma(v)$ is in the core of the game v.*

PROOF. $(i) \Rightarrow (ii) \Rightarrow (iii)$. Trivial.

$(iii) \Rightarrow (iv)$. It is easy to check that $\sum_{j \in N} a_j^\sigma(v) = v(N)$. It suffices to show, therefore, that $\sum_{j \in S} a_j^\sigma(v) \geq v(S)$ for every $S \in \mathcal{N}$. Put

$S := \{j_1, j_2, \ldots, j_s\}$ so that $\sigma(j_1) < \sigma(j_2) < \cdots < \sigma(j_s)$. Apply condition (iii) for $[A = \{j_1, \ldots, j_{i-1}\}, B = P_{j_i}^\sigma, j = j_i]$. Then

$$v(P_{j_i}^\sigma \cup \{j_i\}) - v(P_{j_i}^\sigma) \geq v(\{j_1, \ldots, j_i\}) - v(\{j_1, \ldots, j_{i-1}\}). \qquad (j_i)$$

Summing up the inequalities (j_1)–(j_s), one obtains the required results.

(iv) $\Rightarrow$ (i). Given any $T, S \in \mathcal{N}$, put $T \cap S =: \{j_1, \ldots, j_r\}$; $T \setminus S =: \{j_{r+1}, \ldots, j_s\}$; and $S \setminus T =: \{j_{s+1}, \ldots, j_t\}$, where $0 \leq r \leq s \leq t \leq n$. One may, therefore, put $N \setminus (T \cup S) =: \{j_{t+1}, \ldots, j_n\}$. Now, consider a permutation $\sigma \in G_n$ defined by $\sigma: j_i \mapsto i$. Then $P_{j_i}^\sigma = \{j_1, \ldots, j_{i-1}\}$ for each $i \in N$, so

$$\sum_{j \in S} [v(P_j^\sigma \cup \{j\}) - v(P_j^\sigma)]$$

$$= \sum_{i=s+1}^{t} [v(T \cup \{j_{s+1}, \ldots, j_i\}) - v(T \cup \{j_{s+1}, \ldots, j_{i-1}\})]$$

$$+ \sum_{i=1}^{r} [v(\{j_1, \ldots, j_i\}) - v(\{j_1, \ldots, j_{i-1}\})]$$

$$= [v(T \cup S) - v(T)] + [v(T \cap S) - v(\emptyset)].$$

By condition (iv), this is greater than or equal to $v(S)$. □

Since the core of a side-payment game is a convex polyhedron, the Shapley value of a convex game is always in the core.

6.3. λ-Transfer Value of a Non-Side-Payment Game

Given the set of players N, and hence given the set of nonempty coalitions of players $\mathcal{N}$, a non-side-payment game is defined as a nonempty-valued correspondence $V: \mathcal{N} \to \mathbf{R}^n$ such that for every $S \in \mathcal{N}$, $[u, v \in \mathbf{R}^n, u_i = v_i$ for all $i \in S]$ implies $[u \in V(S)$ iff $v \in V(S)]$. The concept of Shapley value is now extended to this non-side-payment game

6.3. λ-Transfer of a Non-Side-Payment Game

V. For each $\lambda \in \Delta^N$ define a side-payment game $v_\lambda \colon \mathcal{N} \to \mathbf{R}$ by

$$v_\lambda(S) := \sup\{\sum_{j \in S} \lambda_j u_j \mid u \in V(S)\}.$$

The hypothetical scenario behind the concept v_λ is the following: The number λ_j signifies the weight (or the relative significance) of player j in the society, so when he enjoys the utility level u_j, he is regarded as a recipient of the payoff $\lambda_j u_j$ in the society. The payoffs are assumed in this scenario to be freely transferable, hence the side-payment game v_λ.

A utility allocation $u^* \in \mathbf{R}^n$ is called a λ-*transfer value* of V if (1) $u^* \in V(N)$ and (2) there exists $\lambda^* \in \Delta^N$ such that $(\lambda_1^* u_1^*, \ldots, \lambda_n^* u_n^*) = \varphi v_{\lambda^*}$. Condition (2) says that under the weights λ^*, the payoff vector $(\lambda_1^* u_1^*, \ldots, \lambda_n^* u_n^*)$ determined by the utility allocation u^* is the Shapley value of game v_{λ^*}, so it is a fair payoff vector *assuming* transferability of payoffs (i.e., of weighted utilities). Recall the feasibility/efficiency condition (3) of the Shapley value in Section 6.1:

$$\sum_{j \in N} (\varphi v_{\lambda^*})_j = \sup\{\sum_{j \in N} \lambda_j^* u_j \mid u \in V(N)\}.$$

Assuming that sup of the right-hand side is replaced by max there exists $\bar{u} \in V(N)$ such that $\sum_{j \in N} \lambda_j^* u_j^* = \sum_{j \in N} \lambda_j^* \bar{u}_j$. This means that within the framework of the side-payment game v_{λ^*}, the payoff vector $(\lambda_1^* u_1^*, \ldots, \lambda_n^* u_n^*)$, and hence the utility allocation u^*, are obtained by reallocation (i.e., transfer among the members N) of the feasible payoffs $(\lambda_1^* \bar{u}_1, \ldots, \lambda_n^* \bar{u}_n)$. But the scenario for v_{λ^*} is just hypothetical, and transfer of payoffs cannot really take place; the utility allocation u^* may not be feasible. Now condition (1) says that u^* itself, and hence the payoff vector $(\lambda_1^* u_1^*, \ldots, \lambda_n^* u_n^*)$, is feasible via the grand coalition N without any transfer of payoffs. To sum up, a utility allocation u^* is a λ-transfer value if (1) it is feasible via the grand coalition N, and (2) with respect to some weights λ^*, the payoff vector $(\lambda_1^* u_1^*, \ldots, \lambda_n^* u_n^*)$ is precisely the Shapley value (hence a fair outcome) of the hypothetical situation in which an attainable payoff vector $(\lambda_j^* u_j)_{j \in S}$, with $u \in V(S)$, is transferable among the members of S. Recall the notation $\tilde{V}(S) = V(S) \cap \mathbf{R}^S$.

Theorem 6.3.1. *Let* $V: \mathcal{N} \to \mathbf{R}^n$ *be a non-side-payment game. Assume that*

(1) $[S, T \in \mathcal{N}, S \cap T = \varnothing, u \in \tilde{V}(S), v \in \tilde{V}(T)]$ *implies* $[u + v \in V(S \cup T)]$; *and*

(2) *there exists a compact, convex subset* K *of* $\mathbf{R}^n$ *such that* $V(N) = K - \mathbf{R}^n_+$.

Then there exists a λ-transfer value of V. Moreover, any λ-transfer value u^* of V is individually rational, i.e., for the associated weights λ^*,

$$(\varphi v_{\lambda^*})_j \geq \lambda^*_j \sup\{u_j \mid u \in V(\{j\})\}, \quad \textit{for every} \quad j \in N.$$

PROOF. *Step 1.* For each $\lambda \in \Delta^N$, v_λ is a superadditive game; i.e., $v_\lambda(S)$ is a real number for each $S \in \mathcal{N}$, and $v_\lambda(S) + v_\lambda(T) \leq v_\lambda(S \cup T)$ if $[S, T \subset N, \text{ and } S \cap T = \varnothing]$. Indeed, $v_\lambda(N) \in \mathbf{R}$ in view of assumption (2). Choose any $S \in \mathcal{N} \setminus \{N\}$. Then $N \setminus S \in \mathcal{N}$. If there exist $j \in S$ and a sequence $\{u^k\}_k$ in $V(S)$ such that $u^k_j \to \infty$ as $k \to \infty$, choose any $v \in V(N \setminus S)$ and define a sequence $\{w^k\}_k$ by $w^k_i = u^k_i$ if $i \in S$, $w^k_i = v_i$ if $i \in N \setminus S$. By assumption (1), $w^k \in V(N)$ for every k, and yet $w^k_j \to \infty$—a contradiction. Therefore $\tilde{V}(S)$ is bounded from above, and consequently $v_\lambda(S) \in \mathbf{R}$. Superadditivity of v_λ is straightforward in view of assumption (1).

Step 2. For each $\lambda \in \Delta^N$, $(\varphi v_\lambda)_j \geq \lambda_j \sup\{u_j \mid u \in V(\{j\})\}$ for every $j \in N$. Indeed, by formula (††),

$$(\varphi v_\lambda)_j = \frac{1}{n!} \sum_{\sigma \in G_n} [v_\lambda(P^\sigma_j \cup \{j\}) - v_\lambda(P^\sigma_j)].$$

But by superadditivity of v_λ,

$$v_\lambda(P^\sigma_j \cup \{j\}) - v_\lambda(P^\sigma_j) \geq v_\lambda(\{j\})$$

so

$$(\varphi v_\lambda)_j \geq \frac{1}{n!} \sum_{\sigma \in G_n} v_\lambda(\{j\}) = v_\lambda(\{j\}).$$

Step 3. Denote by D_λ the $n \times n$ diagonal matrix whose jth diagonal element is λ_j. Define a correspondence $P: \Delta^N \to \mathbf{R}^n$ by

$$P(\lambda) = \{\pi \in \mathbf{R}^n \mid \sum_{j \in N} \pi_j = 0, \varphi v_\lambda - \pi \in D_\lambda \cdot K\}.$$

6.3. λ-Transfer of a Non-Side-Payment Game

Then for each λ, the set $P(\lambda)$ is nonempty, convex and compact, and the correspondence P is u.s.c. in Δ^N. To show nonemptiness of $P(\lambda)$ notice that

$$\sum_{j \in N} (\varphi v_\lambda)_j = v_\lambda(N) = \sup\{\sum_{j \in N} \lambda_j u_j \mid u \in V(N)\}$$

$$= \max\{\sum_{j \in N} \lambda_j u_j \mid u \in K\}$$

$$=: \sum_{j \in N} \lambda_j \bar{u}_j, \quad \text{with} \quad \bar{u} \in K.$$

It is easy to check that $\varphi v_\lambda - D_\lambda \bar{u} \in P(\lambda)$. The other properties of P are straightforward.

Step 4. Define a correspondence $T: \Delta^N \to \mathbf{R}^n$ by $T(\lambda) := P(\lambda) + \{\lambda\}$. The correspondence T has the same properties as P that are listed in Step 3. By upper semicontinuity of T, one can choose a compact convex subset A of aff $\Delta^N (= \{\alpha \in \mathbf{R}^n \mid \sum_{j \in N} \alpha_j = 1\})$ such that

$$\Delta^N \subset A, \qquad \bigcup_{\lambda \in \Delta^N} T(\lambda) \subset A.$$

Extend T to A by

$$T(\alpha) := T\left(\left(\frac{\max[0, \alpha_j]}{\sum_{i \in N} \max[0, \alpha_i]}\right)_{j \in N}\right).$$

Then the extended correspondence T has Kakutani's fixed point $\alpha^* \in A$. If $\alpha^* \in \Delta^N$, then $\mathbf{0} \in P(\alpha^*)$, so $\varphi v_{\alpha^*} \in D_{\alpha^*} \cdot K$.

Step 5. It suffices to show that $\alpha^* \in \Delta^N$. Assume that $\alpha^* \in A \backslash \Delta^N$, and define $\lambda^* \in \Delta^N$ by $\lambda_j^* = \max[0, \alpha_j^*]/\sum_{i \in N} \max[0, \alpha_i^*]$. Then there exists $k \in N$ such that $\lambda_k^* = 0 > \alpha_k^*$. Since $\alpha^* \in T(\alpha^*) = \{\lambda^*\} + P(\lambda^*)$, there exists $\pi^* \in \mathbf{R}^n$ such that

$$\alpha^* = \lambda^* + \pi^*$$

and

$$\varphi v_{\lambda^*} - \pi^* \in D_{\lambda^*} \cdot V(N).$$

From the first equality $\pi_k^* < 0$; from the second condition $(\varphi v_{\lambda^*})_k - \pi_k^* = 0$. So $(\varphi v_{\lambda^*})_k < 0$. On the other hand, by Step 2 $(\varphi v_{\lambda^*})_k \geq 0$—a contradiction. □

6.4. Value Allocation of a Pure Exchange Economy

Let $\mathscr{E} := \{X^j, \precsim_j, \omega^j\}_{j \in N}$ be a pure exchange economy with n consumers (as formulated in Section 0.2 with the change that $m = n$ here) in which each preference relation $\precsim_j$ is represented by a utility function u^j. As in Section 5.5, one can construct the associated non-side-payment game $V \colon \mathcal{N} \to \mathbf{R}^n$. A *value allocation* of $\mathscr{E}$ is an allocation of commodity bundles $(x^{j*})_{j \in N}$ such that for some choice of a utility function u^j representing $\precsim_j$, hence for some associated non-side-payment game V, $(x^{j*})_{j \in N}$ gives rise to a λ-transfer value of V; it is an n-tuple of commodity bundles $(x^{j*})_{j \in N} \in \prod_{j \in N} X^j$ such that (1) $\sum_{j \in N} x^{j*} \le \sum_{j \in N} \omega^j$ and (2) there exists $\lambda^* \in \Delta^N$ such that $(\lambda_1^* u^1(x^{1*}), \ldots, \lambda_n^* u^n(x^{n*})) = \varphi v_{\lambda^*}$, where $v_{\lambda^*}(S) = \sup\{\sum_{j \in S} \lambda_j^* u_j \mid \exists (x^j)_{j \in S} \in \prod_{j \in S} X^j : \sum_{j \in S} x^j \le \sum_{j \in S} \omega^j$, and $\forall j \in S : u_j \le u^j(x^j)\}$. When $\lambda^* \in \mathrm{icr}\,\Delta^N$, the function $\lambda_j^* u^j(\cdot)$ represents the preference relation $\precsim_j$, so the value allocation in this case has the following interpretation: It is an n-tuple of commodity bundles $(x^{j*})_{j \in N} \in \prod_{j \in N} X^j$ such that (1) it is feasible via the grand coalition N and (2) when one chooses a suitable utility function u^{j*} that represents $\precsim_j$ (more precisely, $u^{j*}(\cdot) := \lambda_j^* u^j(\cdot)$), the resulting utility allocation $(u^{1*}(x^{1*}), \ldots, u^{n*}(x^{n*}))$ is precisely the Shapley value of the associated *side-payment* game $v^* \colon \mathcal{N} \to \mathbf{R}$, in which (newly scaled) utility levels are transferable; $v^*(S) := \sup\{\sum_{j \in S} u_j \mid \exists (x^j)_{j \in S} \in \prod_{j \in S} X^j : \sum_{j \in S} x^j \le \sum_{j \in S} \omega^j$, and $\forall j \in S : u_j \le u^{j*}(x^j)\}$. Notice that the choice of utility functions u^{j*} has two kinds of significance. One significance concerns the ordinal property; the function u^{j*} represents the given preference relation $\precsim_j$. The other significance concerns the cardinal aspect of the way the functions u^{j*} are used; the weight (or the relative significance) of agent j in the economy is determined by the functions $(u^{j*})_{j \in N}$, and the utility levels $(u^{1*}(x^1), \ldots, u^{n*}(x^n))$ are regarded as the payoffs that are freely transferable according to the scenario for game v^*.

Theorem 6.4.1. *Let $\mathscr{E} := \{X^j, \precsim_j, \omega^j\}_{j \in N}$ be a pure exchange economy. Assume for every $j \in N$ that X^j is closed, convex, bounded from below, and $X^j + \mathbf{R}_+^l = X^j$, that $\precsim_j$ is monotone and continuously concavifiable*

6.4. Value Allocation of a Pure Exchange Economy

(*i.e., there exists a continuous, concave function $\bar{u}^j\colon X^j \to \mathbf{R}$ representing $\lesssim_j$), that $\omega^j \in X^j$, and that there exists $c^j \in X^j$ for which $c^j < \omega^j$. Then there exists a value allocation of $\mathscr{E}$. For any value allocation of $\mathscr{E}$, the associated weights λ^* are strictly positive; $\lambda^* \in \mathrm{icr}\,\Delta^N$.*

PROOF. Choose any continuous, concave utility function $\bar{u}^j$ that represents $\lesssim_j$, and construct the associated non-side-payment game $\bar{V}$. Assumption (1) of the preceding theorem is trivially satisfied. To show that assumption (2) is also satisfied, denote by A the set of attainable states $A := \{(x^j)_j \in \prod_{j \in N} X^j \mid \sum_{j \in N} x^j \leq \sum_{j \in N} \omega^j\}$. Since A is compact and the $\bar{u}^j$ are continuous, the set $\bar{u}(A) := \{(\bar{u}^j(x^j))_{j \in N} \in \mathbf{R}^n \mid (x^j)_{j \in N} \in A\}$ is compact; the set $\mathrm{co}\,\bar{u}(A)$ is also compact by Carathéodory's theorem. Clearly, $\bar{V}(N) = \bar{u}(A) - \mathbf{R}_+^n$. It suffices to show, therefore, that $\bar{u}(A) - \mathbf{R}_+^n = \mathrm{co}\,\bar{u}(A) - \mathbf{R}_+^n$. The inclusion "$\subset$" is trivial. Choose any $v \in \mathrm{co}\,\bar{u}(A) - \mathbf{R}_+^n$. Then there exists a finite index set F, and for each $i \in F$ there exist $\alpha_i \in \mathbf{R}$ and $(x^{j,i})_{j \in N} \in A$ such that $\alpha_i \geq 0, \sum_{i \in F} \alpha_i = 1$, and $v \leq \sum_{i \in F} \alpha_i(\bar{u}^1(x^{1,i}), \ldots, \bar{u}^n(x^{n,i}))$. Define $\bar{x}^j := \sum_{i \in F} \alpha_i x^{j,i} \in X^j$. It is easy to verify that $(\bar{x}^j)_{j \in N} \in A$ and $\sum_{i \in F} \alpha_i \bar{u}^j(x^{j,i}) \leq \bar{u}^j(\bar{x}^j)$ by the concavity of $\bar{u}^j$. Thus $v \in \bar{u}(A) - \mathbf{R}_+^n$. All of the assumptions of the preceding theorem are now satisfied, so there exists a λ-transfer value of $\bar{V}$ and hence a value allocation of $\mathscr{E}$.

Finally, let u^j be any utility function representing $\lesssim_j$, $j \in N$. Use $(u^j)_{j \in N}$ to construct an associated non-side-payment game V. Let $(x^{j*})_{j \in N}$ be a value allocation of $\mathscr{E}$ which gives rise to a λ-transfer value of V with the associated weights λ^*. It will be shown that $\lambda^* \gg 0$. Suppose there exists $k \in N$ for which $\lambda_k^* = 0$. Let $S := \{j \in N \mid \lambda_j^* > 0\}$. Then $\varnothing \neq S \subset N \setminus \{k\}$. Now

$$v_{\lambda^*}(S \cup \{k\}) = \sup\{\sum_{j \in S} \lambda_j^* u_j \mid \exists (x^j)_{j \in S \cup \{k\}} \in A(S \cup \{k\}):$$

$$\forall j \in S \cup \{k\} : u_j \leq u^j(x^j)\}$$

$$\geq \sup\{\sum_{j \in S} \lambda_j^* u_j \mid \exists (x^j)_{j \in S} \in A(S):$$

$$\forall j \in S : u_j \leq u^j(x^j + (\omega^k - c^k)/(\#S))\}$$

$$> v_{\lambda^*}(S),$$

where

$$A(T) := \{(x^j)_{j \in T} \in \prod_{j \in T} X^j \mid \sum_{j \in T} x^j \leq \sum_{j \in T} \omega^j\}.$$

Since $[\lambda_k^* = 0]$ also implies $[v_{\lambda*}(\{k\}) = 0]$,

$$v_{\lambda*}(S \cup \{k\}) - v_{\lambda*}(S) > v_{\lambda*}(\{k\}).$$

Notice that assumption (1) of Theorem 6.3.1 holds true even for this game V, so the side-payment game $v_{\lambda*}$ is superadditive. By the same argument as in Step 2 of the proof of Theorem 6.3.1 along with the observation that

$$v_{\lambda*}(P_k^\sigma \cup \{k\}) - v_{\lambda*}(P_k^\sigma) > v_{\lambda*}(\{k\})$$

for all $\sigma \in G_n$ for which $P_k^\sigma = S$, one concludes that $(\varphi v_{\lambda*})_k > v_{\lambda*}(\{k\}) = 0$. This contradicts the definition of λ-transfer value; the latter implies that $0 = \lambda_k^* u^k(x^{k*}) = (\varphi v_{\lambda*})_k$. □

6.5. A Limit Theorem of Value Allocations of Replica Economies

Given a pure exchange economy $\mathscr{E} := \{X^i, \precsim_i, \omega^i\}_{i=1}^m$, denote by $\mathscr{E}^{(q)}$ the q-replica of $\mathscr{E}$ as defined in Section 5.9. In view of the value allocation existence theorem of the preceding section, it will be assumed throughout this section that there exists a continuous, concave function $\bar{u}^i: X^i \to \mathbf{R}$ that represents $\precsim_i$, $i = 1, \ldots, m$, and these $\bar{u}^i$ are used to define the non-side-payment game $\bar{V}^{(q)}$ associated with $\mathscr{E}^{(q)}$:

$$\bar{V}^{(q)}(S) := \{(u_{i,k})_{i=1,k=1}^{m,q} \in \mathbf{R}^{mq} \mid \exists (x^{i,k})_{(i,k) \in S} \in \prod_{(i,k) \in S} X^i : \sum_{(i,k) \in S} x^{i,k}$$

$$\leq \sum_{(i,k) \in S} \omega^i, \text{ and } \forall (i,k) \in S : u_{i,k} \leq \bar{u}(x^{i,k})\}.$$

The purpose of this section is to present one of the classical results of the work done on the *Aumann–Shapley proposition*: In a "large"

6.5. A Limit Theorem of Value Allocations of Replica Economies

economy the set of competitive allocations is characterized as the set of value allocations.

Recall that the formulation in Section 5.9 of a limit theorem of cores (Theorem 5.9.6) is made possible by the equal treatment property of a core allocation of $\mathscr{E}^{(q)}$. One cannot hope for the equal treatment property of every value allocation of $\mathscr{E}^{(q)}$; only certain value allocations having this property will, therefore, be studied here. For this purpose consider the set of weight vectors such that any two agents of the same type are weighted equally:

$$\Delta^{M,(q)} := \{(\lambda_{i,k})_{i=1,k=1}^{m,q} \in \mathbf{R}^{mq} \mid \lambda_{i,k} = \lambda_i \geq 0,$$

$$\text{for } k = 1, \ldots, q, \sum_{i=1}^{m} \lambda_i = 1/q\}.$$

The (i, k) component of $\lambda \in \Delta^{M,(q)}$ will be denoted by λ_i. Given any $\lambda \in \Delta^{M,(q)}$, any two agents of the same type play the same role in the side-payment game $\bar{v}_\lambda^{(q)}$,

$$\bar{v}_\lambda^{(q)}(S) := \sup\{\sum_{(i,k)\in S} \lambda_i u_{i,k} \mid (u_{i,k}) \in \overline{V}^{(q)}(S)\},$$

so by the symmetry axiom, for all $k, k' \in \{1, \ldots, q\}$,

$$(\varphi \bar{v}_\lambda^{(q)})_{i,k} = (\varphi \bar{v}_\lambda^{(q)})_{i,k'} =: (\varphi \bar{v}_\lambda^{(q)})_i.$$

A *symmetric value allocation* of $\mathscr{E}^{(q)}$ is an m-tuple of commodity bundles $(x^{i*})_{i=1}^{m} \in \prod_{i=1}^{m} X^i$ such that (1) $\sum_{i=1}^{m} x^{i*} \leq \sum_{i=1}^{m} \omega^i$ and (2) there exists $\lambda^* \in \Delta^{M,(q)}$ such that

$$(\lambda_1^* \bar{u}^1(x^{1*}), \ldots, \lambda_m^* \bar{u}^m(x^{m*})) = ((\varphi v_{\lambda^*}^{(q)})_1, \ldots, (\varphi v_{\lambda^*}^{(q)})_m).$$

Theorem 6.5.1. *Let* $\mathscr{E} := \{X^i, \lesssim_i, \omega^i\}_{i=1}^{m}$ *be a pure exchange economy, and let* $\mathscr{E}^{(q)}$ *be the q-replica of* $\mathscr{E}$. *Under the same assumptions on* $\mathscr{E}$ *as in the preceding value allocation existence theorem, there exists a symmetric value allocation of* $\mathscr{E}^{(q)}$. *For any symmetric value allocation of* $\mathscr{E}^{(q)}$, *the associated weights* $\lambda^* \in \Delta^{M,(q)}$ *are strictly positive;* $\lambda^* \gg \mathbf{0}$.

PROOF. Repeat the proof of the λ-transfer value existence theorem (Theorem 6.3.1), with $n = mq$, except that the correspondence P:

$\Delta^{M,(q)} \to \mathbf{R}^{mq}$ of Step 3 is now defined as

$$P(\lambda) = \{(\pi_{i,k})_{i,k} \in \mathbf{R}^{mq} \mid \forall k, k' \in \{1, \ldots, q\} : \pi_{i,k} = \pi_{i,k'} = \pi_i.$$

$$\sum_{i=1}^{m} \pi_i = 0. \ \varphi \bar{v}_\lambda^{(q)} - (\pi_{i,k})_{i,k} \in D_\lambda \cdot \overline{K}\},$$

where the set $\overline{K}$ is the convex hull of the attainable utility allocations of $\mathscr{E}^{(q)}$ when the concave utility functions $\bar{u}^i$ are used, and that the compact convex set A of Step 4 is chosen as a subset of aff $\Delta^{M,(q)}$. Thus one obtains a λ-transfer value of $\bar{V}^{(q)}$, $(u^*_{i,k})_{i,k}$, with the associated $\lambda^* \in \Delta^{M,(q)}$. Let $(x^{i,k*})_{i=1,k=1}^{m,q}$ be a commodity allocation which gives rise to $(u^*_{i,k})_{i,k}$. Then by the facts that $\bar{u}^i$ is concave and that

$$\lambda^*_{i,k} = \lambda^*_{i,k'} \quad \text{and} \quad (\varphi \bar{v}^{(q)}_{\lambda^*})_{i,k} = (\varphi \bar{v}^{(q)}_{\lambda^*})_{i,k'}$$

for all $k, k' \in \{1, \ldots, q\}$, it follows that the commodity allocation in which every agent of type i receives $x^{i*} := \sum_{k=1}^{q} x^{i,k*}/q$ is also an allocation for which $\lambda^*_i u^i(x^{i*}) = (\varphi \bar{v}^{(q)}_{\lambda^*})_i$. The last statement of the present theorem is included in the last statement of the preceding value allocation existence theorem. □

One can now state Champsaur's (1975) fundamental contribution to the Aumann–Shapley proposition.

Theorem 6.5.2 (Limit Theorem of Value Allocations). *Let*

$$\mathscr{E} := \{X^i, \lesssim_i, \omega^i\}_{i=1}^{m}$$

be a pure exchange economy, and let $\mathscr{E}^{(q)}$ be the q-replica economy of $\mathscr{E}$. Assume that $\mathscr{E}$ satisfies the following: For each i,

(i) X^i *is closed, convex, bounded from below (there exists $b^i \in \mathbf{R}^l$ such that $X^i \subset \{b^i\} + \mathbf{R}^l_+$) and $X^i + \mathbf{R}^l_+ = X^i$;*

(ii) $\lesssim_i$ *is monotone;*

(iii) *there exists a continuous, strictly concave function $\bar{u}^i : X^i \to \mathbf{R}$ that represents $\lesssim_i$;*

(iv) $\omega^i \in X^i$;

(v) $\sum_{i=1}^{m} \omega^i \in \overset{\circ}{\widehat{\sum_{i=1}^{m} X^i}}$;

6.5. A Limit Theorem of Value Allocations of Replica Economies

(vi) for any $x^i \in X^i$ for which $x^i \gtrsim_i \omega^i$, there exists $y \in \mathbf{R}^l_+ \setminus \{0\}$ such that $x^i - y \in X^i$; and

(vii) for every $p \in \Delta^L : p \cdot \omega^i > \min\{p \cdot x^i \mid x^i \in X^i\}$.

Let $(x^{i,(q)*})_{i=1}^m$ be a symmetric value allocation of $\mathscr{E}^{(q)}$ with the associated weights $\lambda^{(q)*} \in \mathrm{icr}\, \Delta^{M,(q)}$. Set $\theta^{(q)*} := q(\lambda_i^{(q)*})_{i=1}^m$. Let $(\theta^*, (x^{i*})_{i=1}^m)$ be an accumulation point of the sequence $\{(\theta^{(q)*}, (x^{i,(q)*})_{i=1}^m)\}_q$ in the compact set $\{(\theta, (x^i)_{i=1}^m) \in \mathbf{R}^m \times (\mathbf{R}^l)^m \mid 0 \leq \theta_i \leq 1,\ \sum_{i=1}^m \theta_i = 1,\ x^i \in X^i,\ \sum_{i=1}^m x_i \leq \sum_{i=1}^m \omega^i\}$. Then there exists $p^* \in \mathbf{R}^l_+ \setminus \{0\}$ such that $((x^{i*})_{i=1}^m, p^*)$ is a competitive equilibrium of $\mathscr{E}$.

Assumptions (i) and (vii) of the preceding limit theorem together imply assumptions (iv) and (v). Under assumptions (ii) and (vii), assumptions (i) and (iv)–(vi) are all implied by the condition: there exists $b^i \in \mathbf{R}^l$ such that $X^i = \{b^i\} + \mathbf{R}^l_+$. The theorem is stated in the present way, however, in order to clarify the assumptions used in any given step in the proof of the theorem.

The nature of the limit theorem of value allocations (Theorem 6.5.2) is different from the nature of the limit theorem of cores of replica economies (Theorem 5.9.6). To clarify the differences denote by $C(\mathscr{E}^{(q)})$, $V(\mathscr{E}^{(q)})$, and $W(\mathscr{E}^{(q)})$ the set of the core allocations of $\mathscr{E}^{(q)}$, the set of symmetric value allocations of $\mathscr{E}^{(q)}$, and the set of competitive allocations of $\mathscr{E}^{(q)}$, respectively. By the equal treatment property, one may assume that $C(\mathscr{E}^{(q)})$, $V(\mathscr{E}^{(q)})$, and $W(\mathscr{E}^{(q)})$ are all subsets of $\mathbf{R}^{lm}$; in particular, $W(\mathscr{E}^{(q)}) = W(\mathscr{E})$. Now Theorems 5.6.1 and 5.9.6 on core allocations say that (1) $W(\mathscr{E}) \subset C(\mathscr{E}^{(q)})$ for every q; and (2) $W(\mathscr{E}) \supset \bigcap_q C(\mathscr{E}^{(q)})$. The analogy to proposition (1) is no longer true for value allocations; indeed, one can easily construct an example for which $W(\mathscr{E}^{(q)}) \cap V(\mathscr{E}^{(q)}) = \emptyset$. The analogy to proposition (2) also does not hold true for value allocations. Theorem 6.5.2 merely says that the limes superior of the sequence of sets of symmetric value allocations is contained in $W(\mathscr{E})$, i.e., $\mathrm{ls}\{V(\mathscr{E}^{(q)})\}_q \subset W(\mathscr{E})$, where $\mathrm{ls}\{V(\mathscr{E}^{(q)})\}_q := \{x \in \mathbf{R}^{lm} \mid \text{there exists a sequence } \{x^{(q)}\}_q \text{ with } x^{(q)} \in V(\mathscr{E}^{(q)}) \text{ such that } x \text{ is the limit of some subsequence of } \{x^{(q)}\}_q\}$.

Indeed as suggested by Hart (1977), the inclusion may be strict, given the generality of the present assumptions; that is, there may be a competitive allocation in a "large economy" that cannot be characterized as a fair allocation. The other inclusion holds generically true, hence the equivalence of the value allocations and the competitive allocations in a "large economy," if the preference relations are represented by *differentiable* strictly concave utility functions (see Aumann (1975) and Mas-Colell (1977)).

In order to prove the limit theorem (Theorem 6.5.2) several lemmas will be established first. In the first two lemmas an arbitrary positive integer q (hence the q-replica $\mathscr{E}^{(q)}$) is fixed, and the k_0th agent of type i_0, (i_0, k_0), and the coalition S with $S \not\ni (i_0, k_0)$ are also given. Denote by $r_i(S)$ the number of the agents of type i in the coalition S. By abuse of notation define

$$e^i := (0, \ldots, \overset{i}{1}, \ldots, 0) \in \mathbf{R}^m.$$

Then the configuration of S is described by $r(S) := (r_1(S), \ldots, r_m(S)) \in \mathbf{R}^m$; and the configuration of $S \cup \{(i_0, k_0)\}$ by $r(S) + e^{i_0}$. Fix any $\lambda \in \Delta^{M,(q)}$ with $\lambda \gg \mathbf{0}$. The marginal worth of (i_0, k_0)

$$MW^S := \bar{v}_\lambda^{(q)}(S \cup \{(i_0, k_0)\}) - \bar{v}_\lambda^{(q)}(S)$$

will be investigated. By concavity of $\bar{u}^i$, $\bar{v}_\lambda^{(q)}(S)$ is the optimal value of the following:

maximize: $\sum_{i=1}^{m} r_i(S) \lambda_i \bar{u}^i(x^i)$

subject to: $(x^i)_i \in \sum_{i=1}^{m} X^i, \quad \sum_{i=1}^{m} r_i(S) x^i \leq \sum_{i=1}^{m} r_i(S) \omega^i.$

More generally, define

$$D_3(s) := \{\sum_{i=1}^{m} s_i x^i \in \mathbf{R}^l \mid \forall i : x^i \in X^i\}, \quad \text{for each} \quad s \in \overset{\circ}{\mathbf{R}}^m_+;$$

$$D := \{(\lambda, s, z) \in (\text{icr } \Delta^{M,(q)}) \times \overset{\circ}{\mathbf{R}}^m_+ \times \mathbf{R}^l \mid z \in D_3(s)\},$$

and given any $(\lambda, s, z) \in D$, consider the program

maximize: $\sum_{i=1}^{m} s_i \lambda_i \bar{u}^i(x^i),$

subject to: $(x^i)_i \in \prod_{i=1}^{m} X^i, \quad \sum_{i=1}^{m} s_i x^i = z$

[*]

6.5. A Limit Theorem of Value Allocations of Replica Economies

The optimal value $f^s_\lambda(z)$ of $[*]$ is continuous in D and is homogeneous of degree 1 in (s, z) given λ (i.e., $f^{\alpha s}_\lambda(\alpha z) = \alpha f^s_\lambda(z)$ for all $\alpha > 0$); and the concavity of the $\bar{u}^i$ implies that $f^s_\lambda(\cdot)$ is a concave function in $D_3(s)$ so that it has a nonempty superdifferential $\partial f^s_\lambda(z)$ at each point $z \in \overset{\circ}{D}_3(s)$. [Let X be a convex subset of $\mathbf{R}^l$, and let f be a function from X to $\mathbf{R}$. If f is concave, then the subdifferentiability theorem guarantees the existence of a subgradient h of the convex function $-f$ at each point $x \in \overset{\circ}{X}$ (see Exercise 7 of Chapter 1). The linear map $-h$ is called a supergradient of f at x. The superdifferential of f at x is the set of supergradients of f at x.] It is clear that $\partial f^{\alpha s}_\lambda(\alpha z) = \alpha \, \partial f^s_\lambda(z)$ and $\partial f^s_{\alpha\lambda}(z) = \alpha \, \partial f^s_\lambda(z)$. Strict concavity of $\bar{u}^i$ implies that program $[*]$ has a unique solution; in the case of $z = \sum_{i=1}^m s_i \omega^i$, denote the solution by $(x^i(\lambda, s))_{i=1}^m$. It is also clear that $x^i(\lambda, s)$ is homogeneous of degree 0 given λ (i.e., $x^i(\lambda, \alpha s) = x^i(\lambda, s)$ for all $\alpha > 0$).

If $(\lambda, r(S), \sum_{i=1}^m r_i(S)\omega^i) \in D$ and $\sum_{i=1}^m r_i(S)\omega^i \in \overset{\circ}{D}_3(r(S))$, then one can apply the preceding discussion on program $[*]$ to the marginal worth MW^S. It will appear that what matters essentially in the limit theorem is not the absolute configuration $r(S)$ of S but the relative configuration $(1/\sum_{i=1}^m r_i(S)) \cdot r(S)$. So denote by Δ^M the simplex $\mathrm{co}\{e^i \in \mathbf{R}^m \mid i = 1, \ldots, m\}$, and define a function $\tau \colon \overset{\circ}{\mathbf{R}}^m_+ \to \Delta^M$ by

$$\tau(s) := (1/\sum_{i=1}^m s_i) \cdot s.$$

Then using the homogeneities on $f^s_\lambda(z)$,

$$\bar{v}^{(q)}_\lambda(S) = f^{r(S)}_\lambda(\sum_{i=1}^m r_i(S) x^i(\lambda, r(S)))$$

$$= \frac{\#S}{q} f^{\tau(r(S))}_{q\lambda}(\sum_{i=1}^m \tau_i(r(S)) \cdot x^i(\lambda, r(S)))$$

and

$$\bar{v}^{(q)}_\lambda(S \cup \{(i_0, k_0)\}) = \lambda_{i_0} \bar{u}^{i_0}(x^{i_0}(\lambda, r(S) + e^{i_0}))$$

$$+ \sum_{i=1}^m r_i(S) \lambda_i \bar{u}^i(x^i(\lambda, r(S) + e^{i_0})).$$

Here,

$$\sum_{i=1}^{m} r_i(S)\lambda_i \bar{u}^i(x^i(\lambda, r(S) + e^{i_0}))$$

$$\leq f_\lambda^{r(S)}(\sum_{i=1}^{m} r_i(S) \cdot x^i(\lambda, r(S) + e^{i_0}))$$

$$= \frac{\#S}{q} f_{q\lambda}^{\tau(r(S))}(\sum_{i=1}^{m} \tau_i(r(S)) \cdot x^i(\lambda, r(S) + e^{i_0})).$$

So

$$MW^S \leq \lambda_{i_0} \bar{u}^{i_0}(x^{i_0}(\lambda, r(S) + e^{i_0}))$$

$$+ \frac{\#S}{q} [f_{q\lambda}^{\tau(r(S))}(\sum_{i=1}^{m} \tau_i(r(S)) \cdot x^i(\lambda, r(S) + e^{i_0}))$$

$$- f_{q\lambda}^{\tau(r(S))}(\sum_{i=1}^{m} \tau_i(r(S)) \cdot x^i(\lambda, r(S)))].$$

In this case for each $p \in \partial f_{q\lambda}^{\tau(r(S))}(\sum_{i=1}^{m} \tau_i(r(S)) \cdot x^i(\lambda, r(S)))$ the second term of the right-hand side of the preceding inequality is less than or equal to

$$\frac{\#S}{q} p \cdot \sum_{i=1}^{m} (\tau_i(r(S)))(x^i(\lambda, r(S) + e^{i_0}) - x^i(\lambda, r(S)))$$

$$= \frac{1}{q} p \cdot \sum_{i=1}^{m} r_i(S)(x^i(\lambda, r(S) + e^{i_0}) - x^i(\lambda, r(S)))$$

$$= \frac{1}{q} p \cdot (\omega^{i_0} - x^{i_0}(\lambda, r(S) + e^{i_0}))$$

consequently,

$$MW^S \leq \lambda_{i_0} \bar{u}^{i_0}(x^{i_0}(\lambda, r(S) + e^{i_0})) + \frac{1}{q} p \cdot (\omega^{i_0} - x^{i_0}(\lambda, r(S) + e^{i_0})).$$

Denote by e/m the m-dimensional vector $(1/m) \sum_{i=1}^{m} e^i$, each of whose components is $1/m$; this is the relative configuration of a coalition

6.5. A Limit Theorem of Value Allocations of Replica Economies

in which all types have the same number of agents. Clearly, $\sum_{i=1}^{m} (e/m)_i \omega^i \in \overset{\circ}{D}_3(e/m)$. By continuity of $x^i(\ ,\ )$, for any $\theta \in \operatorname{icr} \Delta^M$ there exists a neighborhood $U(\theta, e/m)$ of $(\theta, e/m)$ in $(\operatorname{icr} \Delta^M) \times (\operatorname{icr} \Delta^M)$ such that if $(\lambda, s) \in (\operatorname{icr} \Delta^{M,(q)}) \times \mathbf{R}_+^m$ satisfies $(q(\lambda_i)_{i=1}^m, \tau(s)) \in U(\theta, e/m)$ then $\sum_{i=1}^{m} s_i x^i(\lambda, s) \in \overset{\circ}{D}_3(s)$. Thus

Claim (1) If $(q(\lambda_i)_{i=1}^m, \tau(r(S))) \in U(\theta, e/m)$, then for any

$$p \in \partial f_{q\lambda}^{\tau(r(S))} \left(\sum_{i=1}^{m} \tau_i(r(S)) \cdot x^i(\lambda, r(S)) \right)$$

it follows that

$$q \cdot MW^S \leq (q\lambda_{i_0}) \cdot \bar{u}^{i_0}(x^{i_0}(\lambda, r(S) + e^{i_0})) + p \cdot (\omega^{i_0} - x^{i_0}(\lambda, r(S) + e^{i_0})).$$

Note that in order to establish claim (1), assumptions (i)–(iii) and (v) of the limit theorem (Theorem 6.5.2) have been used.

For notational convenience denote by $(x^i(\theta))_{i=1}^m$ the solution of the following:

$$\text{maximize:} \quad \sum_{i=1}^{m} \theta_i \bar{u}^i(x^i),$$

$$\text{subject to:} \quad (x^i)_i \in \prod_{i=1}^{m} X^i, \quad \sum_{i=1}^{m} x^i = \sum_{i=1}^{m} \omega^i,$$

given $\theta \in \operatorname{icr} \Delta^M$. Notice that $x^i(\lambda, e/m) = x^i(q(\lambda_j)_{j=1}^m)$ for each $\lambda \in \operatorname{icr} \Delta^{M,(q)}$ and that for the value allocation given in the limit theorem, $x^{i,(q)*} = x^i(\theta^{(q)*})$. The result of the following lemma does not depend on the choice of q (as long as the coalition S can be chosen from the agent set for $\mathscr{E}^{(q)}$).

Lemma 6.5.3. *Let $\mathscr{E}$ be a pure exchange economy satisfying assumptions (i)–(iii) and (v) of the limit theorem (Theorem 6.5.2). Choose any $\theta \in \operatorname{icr} \Delta^M$ and any positive number $\epsilon > 0$. Then there exists a neighborhood $U_1(\theta) \times U_2(e/m)$ of $(\theta, e/m)$ in $(\operatorname{icr} \Delta^M) \times (\operatorname{icr} \Delta^M)$, an integer $a(\theta, \epsilon)$, and a price vector $p^* \in \mathbf{R}^l \backslash \{0\}$ such that*

(1) *p^* is a Pareto optimal price vector associated with $(x^i(\theta))_{i=1}^m$; i.e., for each i, $x^i(\theta)$ minimizes the expenditure $p^* \cdot x^i$ on the preferred set $\{x^i \in X^i \mid \bar{u}^i(x^i) \geq \bar{u}^i(x^i(\theta))\}$; and*

(2) if $q(\lambda_i)_i \in U_1(\theta)$ and if $\#S \geq a(\theta, \epsilon)$ and $\tau(r(S)) \in U_2(e/m)$, then

$$q \cdot MW^S \leq (q\lambda_{i_0}) \cdot \bar{u}^{i_0}(x^{i_0}(q(\lambda_i)_i)) + p^* \cdot (\omega^{i_0} - x^{i_0}(\theta)) + \epsilon.$$

PROOF. By the continuity of $x^i(\,,\,)$ on $(\text{icr } \Delta^{M,(q)}) \times (\text{icr } \Delta^M)$, the homogeneity (of degree zero) of $x^i(\lambda, \cdot)$, and the fact that $x^i((\lambda_i)_i, e/m) = x^i(q(\lambda_i)_i)$, it follows that if $(q(\lambda_i)_i, \tau(r(S) + e^{i_0}))$ is close to $(\theta, e/m)$, then $x^{i_0}(\lambda, r(S) + e^{i_0})$ is close to both $x^{i_0}(q(\lambda_i)_i)$ and $x^{i_0}(\theta)$. So there exists a neighborhood $V_1(\theta) \times V_2(e/m)$ of $(\theta, e/m)$ in $(\text{icr } \Delta^M) \times (\text{icr } \Delta^M)$ such that if $(q(\lambda_i)_i, \tau(r(S) + e^{i_0})) \in V_1(\theta) \times V_2(e/m)$, then

(i) $(q\lambda_{i_0}) \cdot \bar{u}^{i_0}(x^{i_0}(\lambda, r(S) + e^{i_0})) + p \cdot (\omega^{i_0} - x^{i_0}(\lambda, r(S) + e^{i_0}))$

$\leq (q\lambda_{i_0}) \cdot \bar{u}^{i_0}(x^{i_0}(q(\lambda_i)_i)) + p \cdot (\omega^{i_0} - x^{i_0}(\theta)) + \epsilon/2$

for all $p \in \Delta^L$.

Note that the correspondence $(\lambda, s) \mapsto \partial f^s_\lambda (\sum_{i=1}^m s_i \omega^i)$ is u.s.c. Consequently, if $(q(\lambda_i)_i, \tau(r(S)))$ is close to $(\theta, e/m)$, then $\partial f^{\tau(r(S))}_{q\lambda}(\sum_{i=1}^m \tau_i(r(S)) \cdot x^i(\lambda, r(S)))$ is contained in a neighborhood of $\partial f^{e/m}_\theta (\sum_{i=1}^m (1/m)\omega^i)$, where θ is a member of $q\Delta^{M,(q)}$ whose (i, k)-component is θ_i for all $k = 1, \ldots, q$. So there exist a neighborhood $W_1(\theta) \times W_2(e/m)$ of $(\theta, e/m)$ in $(\text{icr } \Delta^M) \times (\text{icr } \Delta^M)$ and $p^* \in \partial f^{e/m}_\theta(\sum_{i=1}^m (1/m)\omega^i)$ such that if $(q(\lambda_i)_i, \tau(r(S))) \in W_1(\theta) \times W_2(e/m)$ then:

(ii) $\max\{p \cdot (\omega^{i_0} - x^{i_0}(\theta)) \mid p \in \partial f^{\tau(r(S))}_{q\lambda}(\sum_{i=1}^m \tau_i(r(S)) \cdot x^i(\lambda, r(S)))\}$

$\leq p^* \cdot (\omega^{i_0} - x^{i_0}(\theta)) + \epsilon/2.$

Choose a neighborhood $U_1(\theta) \times U_2(e/m)$ of $(\theta, e/m)$ in $(\text{icr } \Delta^M) \times (\text{icr } \Delta^M)$ such that

$U_1(\theta) \times U_2(e/m)$

$\subset U(\theta, e/m) \cap [V_1(\theta) \times V_2(e/m)] \cap [W_1(\theta) \times W_2(e/m)],$

where $U(\theta, e/m)$ is given in claim (1). Choose this $U_2(e/m)$ small enough and an integer $a(\theta, \epsilon)$ large enough so that $[\#S \geq a(\theta, \epsilon), \tau(r(S)) \in U_2(e/m)]$ implies $[\tau(r(S) + e^{i_0}) \in V_2(e/m)]$. Assertion (2) of the lemma now follows

6.5. A Limit Theorem of Value Allocations of Replica Economies

from claim (1) and the above inequalities (i) and (ii). Trivially, any member of $\partial f_0^{e/m}(\sum_{i=1}^{m}(1/m)\omega^i)$ is a Pareto optimal price vector associated with $(x^i(\theta))_{i=1}^{m}$, which proves assertion (1) of the lemma. □

Lemma 6.5.4. *Let $\mathscr{E}$ be a pure exchange economy satisfying assumptions (i)–(iv) of the limit theorem (Theorem 6.5.2). There exists a positive real number μ such that for any q, any coalition S with $S \not\ni (i_0, k_0)$ in $\mathscr{E}^{(q)}$, and any $\lambda \in \Delta^{M,(q)}$ with $\lambda \gg 0$, it follows that*

$$q \cdot MW^S \leq \mu \cdot (\#S + 1)^2.$$

PROOF. The commodity allocation $(x^i)_{(i,k) \in S \cup \{i_0, k_0\}}$ with the equal treatment property that gives rise to $\bar{v}_\lambda^{(q)}(S \cup \{i_0, k_0\})$ satisfies

$$x^i \leq \sum_{j=1}^{m}(r_j(S) + e_j^{i_0})(\omega^j - b^j) + b^i.$$

So

$$\bar{v}_\lambda^{(q)}(S \cup \{i_0, k_0\}) \leq \sum_{i=1}^{m}(r_i(S) + e_i^{i_0})\lambda_i \bar{u}^i(\sum_{j=1}^{m}(r_j(S) + e_j^{i_0})(\omega^j - b^j) + b^i).$$

On the other hand, it is clear that

$$\bar{v}_\lambda^{(q)}(S) \geq \sum_{i=1}^{m} r_i(S)\lambda_i \bar{u}^i(\omega^i).$$

Therefore

$$MW^S \leq \sum_{i=1}^{m}(r_i(S) + e_i^{i_0})\lambda_i [\bar{u}^i(\sum_{j=1}^{m}(r_j(S) + e_j^{i_0})(\omega^j - b^j) + b^i) - \bar{u}^i(\omega^i)]$$
$$+ \lambda_{i_0} \bar{u}^{i_0}(\omega^{i_0})$$

$$\leq (\#S + 1)\sum_{i=1}^{m}\lambda_i[\bar{u}^i((\#S + 1)\sum_{j=1}^{m}(\omega^j - b^j) + b^i) - \bar{u}^i(\omega^i)]$$
$$+ \lambda_{i_0}\bar{u}^{i_0}(\omega^{i_0}).$$

Note that

$$\frac{1}{\#S+1}\left[(\#S+1)\sum_{j=1}^{m}(\omega^j - b^j) + b^i\right] + \frac{\#S}{\#S+1}\omega^i$$

$$= \sum_{j=1}^{m}(\omega^j - b^j) + \left[\frac{1}{\#S+1}b^i + \frac{\#S}{\#S+1}\omega^i\right],$$

and that this vector is in X^i. So

$$\bar{u}^i((\#S+1)\sum_{j=1}^{m}(\omega^j - b^j) + b^i) + (\#S)\bar{u}^i(\omega^i)$$

$$\leq (\#S+1)\bar{u}^i\left(\sum_{j=1}^{m}(\omega^j - b^j) + \left[\frac{1}{\#S+1}b^i + \frac{\#S}{\#S+1}\omega^i\right]\right)$$

$$\leq (\#S+1)\bar{u}^i(\sum_{j=1}^{m}(\omega^j - b^j) + \omega^i).$$

Thus, in view of the fact that $q(\lambda_i)_i \in \Delta^M$,

$$q \cdot MW^S \leq (\#S+1)\sum_{i=1}^{m}(q\lambda_i)[(\#S+1)\bar{u}^i(\sum_{j=1}^{m}(\omega^j - b^j) + \omega^i)$$

$$- (\#S+1)\bar{u}^i(\omega^i)] + (q\lambda_{i_0})\bar{u}^{i_0}(\omega^{i_0})$$

$$= (\#S+1)^2 \max_i[\bar{u}^i(\sum_{j=1}^{m}(\omega^j - b^j) + \omega^i) - \bar{u}^i(\omega^i)]$$

$$+ \max[0, \max_i \bar{u}^i(\omega^i)].$$

The required result now follows. □

Choose an agent (i_0, k_0) of $\mathscr{E}^{(q)}$ and any integer b such that $1 \leq b \leq mq - 1$. As before, given a permutation σ on the mq agents of $\mathscr{E}^{(q)}$, denote by $P^{\sigma}_{(i_0,k_0)}$ the set of agents in $\mathscr{E}^{(q)}$ who precede (i_0, k_0) with respect to σ. By applying the well-known Chebishev inequality (for the law of large numbers), one could obtain an upper bound of the fraction in the totality (i.e., in the $(mq)!$ permutations on the agents of $\mathscr{E}^{(q)}$) of those permutations σ for which $\#P^{\sigma}_{(i_0,k_0)} = b$ and for which the relative con-

6.5. A Limit Theorem of Value Allocations of Replica Economies

figuration $\tau(r(P^\sigma_{(i_0,k_0)}))$ of the coalition $P^\sigma_{(i_0,k_0)}$ is outside a given neighborhood of e/m. One could obtain, for example, $A/(qb)$ as an upper bound, where A is a number determined by the given neighborhood of e/m. The crucial step in the proof of the limit theorem is the following lemma, which gives a sharper bound for such a fraction. It is based on a strengthened version of the law of large numbers (see the appendix to this section). Choose any positive real number $\delta > 0$, and define

$$\mathcal{S}_i := \{S \subset \{(j,k)\}_{j=1,k=1}^{m,q} \mid |\tau_i(r(S)) - (1/m)| > \delta\}.$$

Given a member S of $\mathcal{S}_i$ for which $\#S = b$, there are $b! \cdot (mq - b - 1)!$ permutations σ such that the first b members with respect to σ are precisely S and $\sigma(i_0, k_0) = b + 1$.

Lemma 6.5.5. *Consider the set $M^{(q)} := \{(i,k) \mid i = 1, \ldots, m, k = 1, \ldots, q\}$ of mq players in which there are q players of type i, $i = 1, \ldots, m$. Choose any positive real number $\delta > 0$, any integer b such that $1 \leq b \leq mq - 1$, and any member $(i_0, k_0) \in M^{(q)}$. Define*

$$\mathcal{S}_i := \{S \subset M^{(q)} \mid \emptyset \neq S \not\ni (i_0,k_0), |\tau_i(r(S)) - (1/m)| \geq \delta\}.$$

Then

$$\sum_{S \in \mathcal{S}_i : \#S = b} \frac{b! \cdot (mq - b - 1)!}{(mq)!} \leq \frac{2}{mq} \exp\left(-\frac{b\delta^2}{2}\right)$$

provided that

$$q > \frac{1}{m} + \frac{2}{\delta m^2} \cdot \max[1, m - 1].$$

PROOF. Consider the following sampling without replacement: The population is $M^{(q)} \setminus \{(i_0, k_0)\}$, from which one draws b members without replacement. Choose any type i. To the tth draw is assigned a variable X_t such that if a player of type i (a player of any other type, resp.) is chosen at the tth draw, then $X_t = 1$ ($X_t = 0$, resp.). Because of random sampling, $(X_1, \ldots, X_b)$ is a random variable—and so is the sample mean $\bar{X} := \sum_{t=1}^{b} X_t/b$. Define

$$E := \begin{cases} (q-1)/(mq-1) & \text{if } i = i_0, \\ q/(mq-1) & \text{if } i \neq i_0. \end{cases}$$

Then by the Hoeffding theorem (see the appendix to this section),

$$\text{prob}\{|\bar{X} - E| \geq \delta/2\} \leq 2\exp\{-b\delta^2/2\}. \tag{i}$$

Now if S is the coalition of players obtained by b draws without replacement and if $(X_1, \ldots, X_b)$ are the associated variables, then $\tau_i(r(S)) = \bar{X}$. Consequently

$$\sum_{S \in \mathscr{S}_i : \#S = b} \frac{b! \cdot (mq - b - 1)!}{(mq - 1)!}$$

$$= \text{prob}\left\{\left|\bar{X} - \left(\frac{1}{m}\right)\right| \geq \delta\right\} \leq \text{prob}\left\{|\bar{X} - E| \geq \frac{\delta}{2}\right\}, \tag{ii}$$

provided that $q > (1/m) + 2\max[1, m - 1]/(\delta m^2)$; the last inequality follows from the fact that under this condition on q, $[|\bar{X} - E| < \delta/2]$ implies $[|\bar{X} - (1/m)| < \delta]$. (See Figure 6.5.1 for the proof of this last fact.) The required result follows from (i) and (ii). □

REMARK. It is worth noting that

$$\sum_{S \in \mathscr{S}_i} \frac{(\#S)! \cdot (mq - \#S - 1)!}{(mq)!} \leq \frac{2}{mq} \sum_{b=0}^{\infty} \left\{\exp\left(-\frac{\delta^2}{2}\right)\right\}^b$$

$$= \frac{2}{mq}\left\{1 - \exp\left(-\frac{\delta^2}{2}\right)\right\}^{-1},$$

so

$$\sum_{S \in \mathscr{S}_1 \cup \cdots \cup \mathscr{S}_m} \frac{(\#S)! \cdot (mq - \#S - 1)!}{(mq)!} \leq \frac{2}{q}\left\{1 - \exp\left(-\frac{\delta^2}{2}\right)\right\}^{-1},$$

provided that $q > (1/m) + 2\max[1, m - 1]/(\delta m^2)$; in particular,

$$\sum_{S \in \mathscr{S}_1 \cup \cdots \cup \mathscr{S}_m} \frac{(\#S)! \cdot (mq - \#S - 1)!}{(mq)!} \to 0 \quad \text{as} \quad q \to \infty.$$

Lemmas 6.5.3–6.5.5 are now used to establish the following main lemma, with which one can complete the proof of the limit theorem straightforwardly.

6.5. A Limit Theorem of Value Allocations of Replica Economies 141

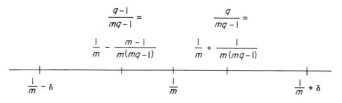

Figure 6.5.1

Lemma 6.5.6. *Let $\mathscr{E}$ be a pure exchange economy satisfying assumptions (i)–(v) of the limit theorem (Theorem 6.5.2). Choose any $\theta \in \mathrm{icr}\,\Delta^M$ and any positive number $\epsilon > 0$. Then there exists a neighborhood $U_1(\theta)$ of θ in $(\mathrm{icr}\,\Delta^M)$, an integer $q(\theta, \epsilon)$, and a price vector $p^* \in \mathbf{R}^l\backslash\{0\}$ such that*

(1) *p^* is a Pareto optimal price vector associated with $(x^i(\theta))_{i=1}^m$; i.e., for each i, $x^i(\theta)$ minimizes the expenditure $p^* \cdot x^i$ on the preferred set $\{x^i \in X^i \mid \bar{u}^i(x^i) \geq \bar{u}^i(x^i(\theta))\}$; and*

(2) *if $q(\lambda_i)_i \in U_1(\theta)$, and if $q \geq q(\theta, \epsilon)$, then for every $i \in \{1, \ldots, m\}$,*

$$q \cdot (\varphi v_\lambda^{(q)})_i \leq q\lambda_i \bar{u}^i(x^i(q(\lambda_j)_j)) + p^* \cdot (\omega^i - x^i(\theta)) + 4\epsilon.$$

PROOF. The notation introduced in the statements of Lemmas 6.5.3, 6.5.4, and 6.5.5 will be freely used here. Choose any $(i_0, k_0) \in M^{(q)}$, and define:

$$\mathscr{C}_1 := \{S \subset M^{(q)} \mid S \not\ni (i_0, k_0).\ \#S \geq a(\theta, \epsilon).\ \tau(r(S)) \in U_2(e/m)\},$$
$$\mathscr{C}_2 := \{S \subset M^{(q)} \mid S \not\ni (i_0, k_0).\ \#S < a(\theta, \epsilon).\ \tau(r(S)) \in U_2(e/m)\},$$
$$\mathscr{C}_3 := \{S \subset M^{(q)} \mid S \not\ni (i_0, k_0).\ \tau(r(S)) \notin U_2(e/m)\},$$

and

$$\gamma_S := (\#S)! \cdot (mq - \#S - 1)!/[(mq)!].$$

It is easy to check that $\{\mathscr{C}_1, \mathscr{C}_2, \mathscr{C}_3\}$ is a partition of the power set of $M^{(q)}\backslash\{(i_0, k_0)\}$ and that

$$\sum_{S \in \mathscr{C}_2} \gamma_S = \frac{1}{mq} \sum_{b=0}^{a(\theta,\epsilon)-1} \sum_{S \in \mathscr{C}_2 : \#S = b} \frac{b!(mq - 1 - b)!}{(mq - 1)!} \leq \frac{a(\theta, \epsilon)}{mq}.$$

Now by Lemma 6.5.3

$$\sum_{S \in \mathscr{C}_1} \gamma_S q(MW^S)$$

$$\leq \sum_{S \in \mathscr{C}_1} \gamma_S \cdot \{q\lambda_{i_0}\bar{u}^{i_0}(x^{i_0}(q(\lambda_i)_i)) + p^* \cdot (\omega^{i_0} - x^{i_0}(\theta)) + \epsilon\}$$

$$\leq \{\sum_{S \in \mathscr{C}_1} \gamma_S\} \cdot \{q\lambda_{i_0}\bar{u}^{i_0}(x^{i_0}(q(\lambda_i)_i)) + p^* \cdot (\omega^{i_0} - x^{i_0}(\theta)) + \epsilon\}. \quad \text{(i)}$$

By Lemma (6.5.4),

$$\sum_{S \in \mathscr{C}_2} \gamma_S q(MW^S) \leq \mu(a(\theta, \epsilon) + 1)^2 \frac{a(\theta, \epsilon)}{mq} \leq \epsilon \quad \text{(ii)}$$

for all q sufficiently large, say, for all $q \geq q_2(\theta, \epsilon)$. Choose the positive number δ of Lemma 6.5.5 small enough so that $[S \in \mathscr{C}_3]$ implies $[S \in \mathscr{S}_1 \cup \cdots \cup \mathscr{S}_m]$. Then by Lemmas 6.5.4 and 6.5.5,

$$\sum_{S \in \mathscr{C}_3} \gamma_S q(MW^S)$$

$$\leq \sum_{b=0}^{\infty} \sum_{S \in \mathscr{S}_1 \cup \cdots \cup \mathscr{S}_m, \#S = b} \frac{b! \cdot (mq - b - 1)!}{(mq)!} \cdot q(MW^S)$$

$$\leq \sum_{b=0}^{\infty} \frac{2}{q} \left\{ \exp\left(-\frac{b\delta^2}{2}\right) \right\} \mu(b + 1)^2$$

$$\leq \epsilon \quad \text{(iii)}$$

for all q sufficiently large; say for all $q \geq q_3(\theta, \epsilon)$. Finally,

$$1 \geq \sum_{S \in \mathscr{C}_1} \gamma_S = 1 - \sum_{S \in \mathscr{C}_2} \gamma_S - \sum_{S \in \mathscr{C}_3} \gamma_S$$

$$\geq 1 - \frac{a(\theta, \epsilon)}{mq} - \frac{2}{q} \left\{ 1 - \exp\left(-\frac{\delta^2}{2}\right) \right\}^{-1}$$

$$\to 1 \quad \text{as} \quad q \to \infty. \quad \text{(iv)}$$

In view of the fact that

$$q(\varphi v^{(q)})_{i_0} = \sum_{S \in \mathscr{C}_1 \cup \mathscr{C}_2 \cup \mathscr{C}_3} \gamma_S q(MW^S),$$

the required result follows from (i)–(iv). □

6.5. A Limit Theorem of Value Allocations of Replica Economies

PROOF OF THE LIMIT THEOREM (THEOREM 6.5.2). *Step 1.* $\theta^* \gg 0$. Indeed since $x^{i,(q)*} = x^i(\theta^{(q)*})$ for every q, it follows from the maximum theorem that $x^{i*} = x^i(\theta^*)$, i.e., $(x^{i*})_i$ is a solution of

$$\text{maximize:} \quad \sum_{i=1}^{m} \theta_i^* \bar{u}^i(x^i),$$

$$\text{subject to:} \quad (x^i)_i \in \prod_{i=1}^{m} X^i, \quad \sum_{i=1}^{m} x^i = \sum_{i=1}^{m} \omega^i.$$

Suppose there exists k such that $\theta_k^* = 0$. Then $\bar{u}^k(x^{k*}) < \bar{u}^k(\omega^k)$; for otherwise there would exist $y \in \mathbf{R}_+^l \setminus \{0\}$ such that $x^{k*} - (m-1)y \in X^k$ by assumption (vi), so

$$\sum_{i \neq k} \theta_i^* \bar{u}^i(x^{i*} + y) + \theta_k^* \bar{u}^k(x^{k*} - (m-1)y) > \sum_{i=1}^{m} \theta_i^* \bar{u}^i(x^{i*})$$

by assumption (ii), which contradicts the optimality of $(x^{i*})_i$. On the other hand, by the individual rationality of the value allocation $\bar{u}^k(x^{k,(q)*}) \geq \bar{u}^k(\omega^k)$ for all q, so $\bar{u}^k(x^{k*}) \geq \bar{u}^k(\omega^k)$—a contradiction.

Step 2. Apply Lemma 6.5.6 with $\theta = \theta^*$. Then there exists a neighborhood $U_1(\theta^*)$ of θ^* in (icr Δ^M), an integer $q(\theta^*, \epsilon)$, and a price vector $p^* \in \mathbf{R}^l \setminus \{0\}$ such that

(1) p^* is a Pareto optimal price vector associated with $(x^{i*})_{i=1}^m$ and
(2) for all $q \geq q(\theta^*, \epsilon)$ for which $\theta^{(q)*} \in U_1(\theta^*)$, it follows that for each i,

$$q \cdot (\varphi v_{\lambda^{(q)*}}^{(q)})_i \leq \theta_i^{(q)*} \bar{u}^i(x^{i,(q)*}) + p^* \cdot (\omega^i - x^{i*}) + 4\epsilon.$$

Since $(x^{i,(q)*})_i$ is a symmetric value allocation with associated weights $\lambda^{(q)*}$, property (2) implies that

$$0 \leq p^* \cdot (\omega^i - x^{i*}) + 4\epsilon,$$

which is true for any $\epsilon > 0$ so

$$0 \leq p^* \cdot (\omega^i - x^{i*}) \quad \text{for every} \quad i.$$

Now it is clear that $0 = \sum_{i=1}^{m} (\omega^i - x^{i*})$ so that $0 = p^* \cdot \sum_{i=1}^{m} (\omega_i - x^{i*})$. Thus

$$p^* \cdot \omega^i = p^* \cdot x^{i*}, \quad \text{for every} \quad i.$$

In view of assumption (vii), one can now apply the same argument as that in Step 3 of the proof of the second fundamental theorem of welfare economics (Theorem 4.5.2) to establish that x^{i*} is a maximal element of the budget set $\{x^i \in X^i \,|\, p^* \cdot x^i \leq p^* \cdot \omega^i\}$ with respect to $\lesssim_i$.

The monotonicity assumption on $\lesssim_i$ implies that $p^* \gg 0$. $\square$

APPENDIX TO SECTION 6.5: THE LAW OF LARGE NUMBERS

Let $X_1, \ldots, X_b$ be independent random variables with the common finite expectation E and the common finite variance V^2. Denote by $\bar{X}^{(b)}$ the arithmetic mean $(X_1 + \cdots + X_b)/b$. Then the random variable $\bar{X}^{(b)}$ has an expectation E and a variance V^2/b. Chebishev's inequality may be stated as: For any positive real number $\delta > 0$,

$$\text{prob}\{|\bar{X}^{(b)} - E| \geq \delta\} \leq V^2/(b\delta^2).$$

The law of large numbers follows from this inequality since $V^2/(b\delta^2) \to 0$ as $b \to \infty$.

Hoeffding (1963, Theorem 1) replaced the term $V^2/(b\delta^2)$ by several numbers and obtained sharper results. In particular, if $0 \leq X_t \leq 1$ for all $t = 1, \ldots, b$,

$$\text{prob}\{|\bar{X}^{(b)} - E| \geq \delta\} \leq 2 \exp\{-2b\delta^2\}.$$

One can obtain a similar result for a random sampling *without* replacement from a finite population. Let N be a population with $\#N = n$. A real number c_j is associated with each member $j \in N$ such that $0 \leq c_j \leq 1$. Define

$$E := \sum_{j \in N} c_j/n,$$

and choose any positive integer b with $b \leq n$. The random sampling without replacement is now described by $(X_1, \ldots, X_b)$, where $X_t = c_j$ if j is obtained at the tth draw. Set $\bar{X}^{(b)} := (X_1 + \cdots + X_b)/b$. Hoeffding (1963, Section 6) also established that

$$\text{prob}\{|\bar{X}^{(b)} - E| \geq \delta\} \leq 2 \exp\{-2b\delta^2\}.$$

6.6. Notes

The definition of the Shapley value φ and the theorem (Theorem 6.1.1) for its existence and uniqueness are due to Shapley (1953), although the role of null players in his original treatment was implicit. The present proof of Theorem 6.1.1 is taken from Aumann and Shapley (1974, Appendix A). See also Dubey (1975) for a proof of the uniqueness of the Shapley value.

Other characterizations of the Shapley value have been established: Roth (1977) formulated a utility of playing a game and established a condition under which the utility function is precisely the Shapley value. Myerson (1977) formulated communication networks connecting the players, proposed an allocation rule for a game v that assigns a payoff distribution among the players to each network, and proved that the rule is an extension of the Shapley value.

There are several ways to define the Shapley value, given infinitely many players, in particular, given an atomless probability measure space of players $(A, \mathscr{A}, v)$. The following are two classical ways to do so: Recall that **ba** denotes the vector space of all bounded additive scalar functions defined on $\mathscr{A}$. Aumann and Shapley (1974) identified vector spaces of games on which there exists a function to **ba** satisfying (analog of) the four axioms of Section 6.1; such a function is called the *value*. Kannai's (1966) approach is, on the other hand, based on formula (†): Given a game $v: \mathscr{A} \to \mathbf{R}$ and a coalition $S \in \mathscr{A}$, consider an $\mathscr{A}$-measurable, finite partition Π of A such that S is the union of some members of Π. Regard a member of Π as a "player," and construct a game v_Π having the finite player set Π by defining

$$v_\Pi(\Xi) := v(\bigcup_{T \in \Xi} T), \quad \text{for each } \Xi \subset \Pi.$$

One obtains the Shapley value φv_Π by using (†), and hence $\sum_{T \in \Pi: T \subset S} (\varphi v_\Pi)_T$. When this last term converges to a real number $(\varphi v)(S)$, as the partition Π becomes finer, and when the number $(\varphi v)(S)$ does not depend upon the choice of a sequence of partitions $\{\Pi\}_\Pi$, the set function φv is called the *asymptotic value* of v. The study of (Shapley) values of games with infinitely many players would probably be the deepest mathematical study among all the social sciences; a brief survey of its recent developments can be found in Aumann (1980).

The concept of convex games was proposed and studied by Shapley (1971), who established the statement $[(i) \Leftrightarrow (ii) \Leftrightarrow (iii) \Rightarrow (iv)]$ of theorem 6.2.1. Under condition (iii), the marginal worth vectors are precisely the vertices of the core of v. Ichiishi (1981c) pointed out that $[(iv) \Rightarrow (i)]$. R. Weber (1978) established that given any side-payment game, its core is contained in the convex hull of the set of marginal worth vectors (see Exercise 1 of this Chapter). A related work is found in Tijs (1981), who proposed a new concept: the *τ-value*. For the (Shapley) value of convex games with infinitely many players see Rosenmüller (1971).

The explicit formulation of the concept of λ-transfer value and its existence theorem (Theorem 6.3.1) are due to Shapley (1969). There are several predecessors leading to this concept; in particular, see Harsanyi (1963). Roth (1980) constructed an example of the three-person non-side-payment game for which the λ-transfer value yields a result difficult to justify. Exercise 2 adapts Jean-Pierre Aubin's technique for nonemptiness of the core under the strong balancedness condition. Owen (1972) proposed an alternative concept of the value of a non-side-payment game, using his earlier idea of multilinear extension of the Shapley value. A λ-transfer value existence theorem, given infinitely many players, was established by Neyman (1977).

The definition of value allocations of a pure exchange economy $\mathscr{E} := \{X^j, \lesssim_j, \omega^j\}_{j \in N}$, as given in Section 6.4, is based on the choice of a utility function u^j that represents $\lesssim_j$, and concavity of such a utility function is needed in the existence theorem (Theorem 6.4.1). Some convex preference relations cannot, however, be represented by concave utility functions even if they are complete, transitive, and closed (hence represented by quasi-concave, continuous functions). Moreover, there is an economy with such preference relations such that the associated set of attainable utility allocations $V(N)$ cannot be convex no matter which utility functions are used to represent the given preference relations [See Kannai and Mantel (1978)]. Shafer (1980) called the value allocation of Section 6.4 *cardinal* and distinguished it from another value allocation concept that he called *ordinal*; the latter concept had

6.6. Notes

been used by Aumann (1975). An *ordinal value allocation* of $\mathscr{E}$ is a commodity allocation $(x^{j*})_{j \in N} \in \prod_{j \in N} X^j$ such that (1) $\sum_{j \in N} x^{j*} = \sum_{j \in N} \omega^j$; and (2) there exists a utility function u^{j*} representing $\lesssim_j$ for each j such that $(u^{j*}(x^{j*}))_{j \in N}$ is the Shapley value of game v^* defined by $v^*(S) := \sup\{\sum_{j \in S} u_j \mid \exists (x^j)_{j \in S} \in \prod_{j \in S} X^j : \sum_{j \in S} x^j \leq \sum_{j \in S} \omega^j$, and $\forall j \in S : u_j \leq u^{j*}(x^j)\}$. (Recall the discussion in Section 6.4 of the cardinal value allocation with $\lambda^* \gg \mathbf{0}$.) Shafer (1980) then established an ordinal value allocation existence theorem for an economy in which each preference relation satisfies merely completeness, transitivity, closedness, monotonicity, and weak convexity. He also presented an insightful discussion of the interpretation of value allocation.

The Aumann–Shapley proposition that value allocations characterize competitive allocations in a "large" economy goes back to Shapley (1964). Given a pure exchange economy $\mathscr{E} := \{X^i, \lesssim_i, \omega^i\}_{i=1}^m$, an economy with transferable utility is constructed as follows: Let u^i be a utility function that represents $\lesssim_i$. One introduces the $(l+1)$st commodity, "money," whose price is always equal to 1. The new consumption set of consumer i becomes $X^i \times \mathbf{R}$ and his new preference relation is represented by a function $(x^i, x^i_{l+1}) \mapsto u^i(x^i) + x^i_{l+1}$. Given a price vector $p \in \Delta^L$, his problem becomes

maximize: $u^i(x^i) + x^i_{l+1}$
subject to: $(p, 1) \cdot (x^i, x^i_{l+1}) = (p, 1) \cdot (\omega^i, 0)$, $(x^i, x^i_{l+1}) \in X^i \times \mathbf{R}$.

The market clearance condition becomes

$$\sum_i (x^i, x^i_{l+1}) \leq \sum_i (\omega^i, 0).$$

The cooperative game in characteristic function form induced by an economy with transferable utility (see Section 5.5) is a side-payment game. Its value allocation is, therefore, a commodity allocation that gives rise to the Shapley value of the game. Now Shapley (1964) established a limit theorem of the value allocations of replica economies with transferable utility. Based on their value theory and on a variational problem of Aumann and Perles (1965), Aumann and Shapley (1974) established a value equivalence theorem for an atomless economy with transferable, differentiable utility. This work serves as important stepping stones in establishing the Aumann–Shapley proposition for economies *without* transferable utility (e.g., for pure exchange economies). Theorem 6.5.2 and its proof in this text are due to Champsaur (1975). Lemma 6.5.5 is taken from Cheng (1981). Aumann (1975) established an equivalence theorem for an atomless pure exchange economy with "uniformly smooth" utility. See also Hart (1977) and Mas-Colell (1977). Owen (1976) considered a limit theorem for the value concept as defined in Owen (1972).

EXERCISES

1. Denote by G^N the set of all side-payment games with the player set $N := \{1, \ldots, n\}$. Let G_n be the symmetric group on N, and for each $v \in G^N$ and $\sigma \in G_n$ denote by $a^\sigma(v)$ the marginal worth vector; denote by $C(v)$ the core. Prove that $C(v) \subset \mathrm{co}\{a^\sigma(v) \mid \sigma \in G_n\}$ for every $v \in G^N$. (*Hint*: By induction on n. It suffices to show that $u \in C(v)\backslash\mathrm{icr}\, C(v) \Rightarrow u \in \mathrm{co}\{a^\sigma(v) \mid \sigma \in G_n\}$. $[u \in C(v)\backslash\mathrm{icr}\, C(v)]$ implies $\exists S \in \mathcal{N}\backslash\{N\} : \sum_{j \in S} u_j = v(S)$. Consider games $v' \in G^S$ and $v'' \in G^{N\backslash S}$ defined by

$$v'(T) := v(T) \qquad \text{for all} \quad T \subset S$$
$$v''(T) := v(S \cup T) - v(S) \qquad \text{for all} \quad T \subset N\backslash S.$$

Then $(u_j)_{j \in S} \in C(v')$ and $(u_j)_{j \in N\backslash S} \in C(v'')$.)

2. Let $V: \mathcal{N} \to \mathbf{R}^n$ be a non-side-payment game. Assume that (1) $\forall S, T \in \mathcal{N} : \tilde{V}(S) + \tilde{V}(T) \subset \tilde{V}(S \cap T) + \tilde{V}(S \cup T)$ and (2) there exists a compact, convex subset K of $\mathbf{R}^n$ such that $V(N) = K - \mathbf{R}^n_+$.

   (i) Define for each $\lambda \in \Delta^N$,

   $$v_\lambda(S) := \sup\{\sum_{j \in S} \lambda_j u_j \mid u \in \tilde{V}(S)\}.$$

   Show that v_λ is a convex game.

   (ii) Explain briefly why the map $\Delta^N \to \mathbf{R}^n$, $\lambda \mapsto \varphi v_\lambda$, is continuous, and for each $\lambda \in \Delta^N$, φv_λ is in the core of v_λ.

   (iii) For each positive integer v, define

   $$\Delta^N_v := \{\lambda \in \Delta^N \mid \forall j \in N : \lambda_j \geq 1/v\}$$

Exercises 149

For each $\lambda \in \Delta^N$, denote by D_λ the $n \times n$ diagonal matrix whose jth diagonal element is λ_j. Show that there exists a compact convex subset E_v of $\mathbf{R}^n$ such that

$$D_\lambda^{-1} \cdot (\varphi v_\lambda) \in E_v \quad \text{for every} \quad \lambda \in \Delta_v^N.$$

(iv) Define a correspondence $G_v : \mathbf{R}^n \to \Delta_v^N$ by: $G_v(u)$ is the set of solutions of

$$\begin{aligned} \text{minimize:} & \quad v_\lambda(N) - \sum_{j \in N} \lambda_j u_j \\ \lambda & \\ \text{subject to:} & \quad \lambda \in \Delta_v^N. \end{aligned}$$

Show that G_v is u.s.c., non-empty-valued, convex-valued, and closed-valued for any given $v \geq n$.

Consider a map $\Delta_v^N \times E_v \to \Delta_v^N \times E_v, (\lambda, u) \mapsto G_v(u) \times \{D_\lambda^{-1} \cdot (\varphi v_\lambda)\}$. Let (λ^v, u^v) be Kakutani's fixed-point.

(v) Fix $v \geq n$, and prove that

$$\forall S \in \mathcal{N} : u^v \in \overline{\mathbf{R}^n \backslash V(S)},$$

$$\forall \lambda \in \Delta_v^N : \sum_{j \in N} \lambda_j u_j^v \leq v_\lambda(N).$$

(vi) Prove for any $v \geq n$,

$$u_j^v \geq \sup\{u_j \in \mathbf{R} \mid u \in V(\{j\})\} \quad \text{for all} \quad j \in N,$$

$$\sum_{j \in N} u_j^v \leq n v_{(1/n)\chi_N}(N).$$

(vii) Let (λ^*, u^*) be an accumulation point of $\{\lambda^v, u^v\}_v$. Prove that

$$D_{\lambda^*} u^* = \varphi v_{\lambda^*}, \quad \forall S \in \mathcal{N} : u^* \in \overline{\mathbf{R}^n \backslash V(S)}, \quad \text{and} \quad u^* \in V(N).$$

Interpret these results.

References

Anderson, R. M. (1978). An elementary core equivalence theorem. *Econometrica* **46**, 1483–1487.

Anderson, R. M. (1981). Core theory with strongly convex preferences. *Econometrica* **49**, 1457–1468.

Arrow, K. J. (1951). An extension of the basic theorems of classical welfare economics. In J. Neyman (Ed.), *Proc. Berkeley Symp. Math. Statistics and Probability*, 2nd, pp. 507–532. Berkeley: Univ. of California Press.

Arrow, K. J., and Debreu, G. (1954). Existence of an equilibrium for a competitive economy. *Econometrica* **22**, 265–290.

Arrow, K. J., and Hahn, F. H. (1971). *General Competitive Analysis*. San Francisco: Holden-Day.

Artstein, Z. (1980). Discrete and continuous bang-bang and facial spaces or: Look for the extreme points. *SIAM Rev.* **22**, 172–185.

Aubin, J.-P. (1979). *Mathematical Methods of Game and Economic Theory*. Amsterdam: North-Holland.

Aumann, R. J. (1959). Acceptable points in general cooperative n-person games. In A. W. Tucker and R. D. Luce (Eds.), *Contributions to the Theory of Games*, Vol. IV, pp. 287–324. Princeton: Princeton Univ. Press.

Aumann, R. J. (1961). The core of a cooperative game without side payments. *Trans. Amer. Math. Soc.* **98**, 539–552.

Aumann, R. J. (1964). Markets with a continuum of traders. *Econometrica* **32**, 39–50.

Aumann, R. J. (1966). Existence of competitive equilibria in markets with a continuum of traders. *Econometrica* **34**, 1–17.

Aumann, R. J. (1975). Values of markets with a continuum of traders. *Econometrica* **43**, 611–646.

Aumann, R. J. (1980). Recent developments in the theory of the Shapley value. In O. Lehto (Ed.), *Proc. Internat. Congr. Mathematicians, Helsinki, 1978*, pp. 995–1003. Hungary: Academia Scientiarum Fennica.

Aumann, R. J., and Peleg, B. (1960). Von Neumann–Morgenstern solutions to cooperative games without side payments. *Bull. Amer. Math. Soc.* **66**, 173–179.

Aumann, R. J., and Perles, M. (1965). A variational problem arising in economics. *J. Math. Anal. Appl.* **11**, 488–503.

Aumann, R. J., and Shapley, L. S. (1974). *Values of Non-Atomic Games*. Princeton: Princeton Univ. Press.

Berge, C. (1959). *Espaces Topologiques*. Paris: Dunod. (English transl.: *Topological Spaces*. Edinburgh and London: Oliver and Boyd, 1963.)

Bewley, T. (1973). Edgeworth's conjecture. *Econometrica* **41**, 415–452.

Billera, L. J. (1970). Some theorems on the core of an n-person game without side-payments. *SIAM J. Appl. Math.* **18**, 567–579.

Billera, L. J. (1974). On games without side payments arising from a general class of markets. *J. Math. Econom.* **1**, 129–139.

Bondareva, O. N. (1962). Teoriia iadra v igre n lits. *Vestnik Leningrad Univ. Math.* **13**, 141–142.

Bondareva, O. N. (1963). Nekotorye primeneniia metodov linejnogo programmirovaniia k teorii kooperativnykh igr. *Problemy Kibernet.* **10**, 119–139.

Brouwer, L. E. J. (1012). Über Abbildungen von Mannifaltigkeiten. *Math. Ann.* **71**, 97–115.

Browder, F. E. (1967). A new generalization of the Schauder fixed point theorem. *Math. Ann.* **174**, 285–290.

Cellina, A. (1969). Approximation of set valued functions and fixed point theorems. *Ann. Mat. Pura Appl.* **82**, 3–24.

Champsaur, P. (1975). Cooperation versus competition. *J. Econom. Theory* **11**, 394–417.

Cheng, H.-C. (1981). On dual regularity and value convergence theorems. *J. Math. Econom.* **8**, 37–57.

Cournot, A. A. (1838). *Recherches sur les principes mathématiques de la théorie des richesses*, Paris: Libraire des sciences politiques et sociales, M. Rivière & cie. (English transl.: *Researches into the Mathematical Principles of the Theory of Wealth*. New York: Macmillan, 1897.)

Debreu, G. (1952). A social equilibrium existence theorem. *Proc. Nat. Acad. Sci. U.S.A.* **38**, 886–893.

Debreu, G. (1956). Market equilibrium. *Proc. Nat. Acad. Sci. U.S.A.* **42**, 876–878.

Debreu, G. (1959). *Theory of Value*. New York: Wiley.

Debreu, G. (1970). Economies with a finite set of equilibria. *Econometrica* **38**, 387–392.

Debreu, G., and Scarf, H. (1963). A limit theorem on the core of an economy. *Internat. Econom. Rev.* **4**, 235–246.

Debreu, G., and Scarf, H. (1972). The limit of the core of an economy. In C. B. McGuire and R. Radner (Eds.), *Decision and Organization*, pp. 283–295. Amsterdam: North-Holland.

Delbaen, F. (1974). Convex games and extreme points. *J. Math. Anal. Appl.* **45**, 210–233.

Dubey, P. (1975). On the uniqueness of the Shapley value. *Internat. J. Game Theory* **4**, 131–139.

Edgeworth, F. Y. (1881). *Mathematical Psychics*. London: Kegan Paul.

References

Fan, K. (1952). Fixed-point and minimax theorems in locally convex topological linear spaces. *Proc. Nat. Acad. Sci. U.S.A.* **38**, 121–126.

Fan, K. (1956). On systems of linear inequalities. In H. W. Kuhn and A. W. Tucker (Eds.), *Linear Inequalities and Related Systems*, pp. 99–156. Princeton: Princeton Univ. Press.

Fan, K. (1972). A minimax inequality and applications. In O. Shisha (Ed.), *Inequalities III*, pp. 103–113. New York: Academic Press.

Fenchel, W. (1951). *Convex Cones, Sets and Functions*. Lecture Notes. Princeton Univ., Princeton.

Gale, D. (1955). The law of supply and demand. *Math. Scand.* **3**, 155–169.

Gillies, D. B. (1959). Solutions to general non-zero-sum games. In A. W. Tucker and R. D. Luce (Eds.), *Contributions to the Theory of Games*, Vol. IV, pp. 47–85. Princeton: Princeton Univ. Press.

Glicksberg, I. L. (1952). A further generalization of the Kakutani fixed point theorem, with application to Nash equilibrium points. *Proc. Amer. Math. Soc.* **3**, 170–174.

Harsanyi, J. C. (1963). A simplified bargaining model for the n-person cooperative game. *Internat. Econom. Rev.* **4**, 194–220.

Hart, S. (1977). Values of non-differentiable markets with a continuum of traders. *J. Math. Econom.* **4**, 103–116.

Hausdorff, F. (1927). *Mengenlehre*. Berlin: W. de Gruyter. (English transl.: *Set Theory*. New York: Chelsea, 1962.)

Hildenbrand, W. (1970). On economies with many agents. *J. Econom. Theory* **2**, 161–188.

Hildenbrand, W. (1974). *Core and Equilibria of a Large Economy*. Princeton: Princeton Univ. Press.

Hoeffding, W. (1963). Probability inequalities for sums of bounded random variables. *J. Amer. Statist. Assoc.* **58**, 13–30.

Howe, R. (1979). On the tendency toward convexity of the vector sum of sets. Cowles Foundation Discussion Paper No. 538, Yale Univ., New Haven.

Ichiishi, T. (1981a). On the Knaster–Kuratowski–Mazurkiewicz–Shapley theorem. *J. Math. Anal. Appl.* **81**, 297–299.

Ichiishi, T. (1981b). A social coalitional equilibrium existence lemma. *Econometrica* **49**, 369–377.

Ichiishi, T. (1981c). Super-modularity: Applications to convex games and to the greedy algorithm for LP. *J. Econom. Theory* **25**, 283–286.

Ichiishi, T. (1982a). Non-cooperation and cooperation. In M. Deistler, E. Fürst, and G. Schwödiauer (Eds.), *Games, Economic Dynamics, and Time Series Analysis*, pp. 14–48. Vienna/Würzburg: Physica-Verlag.

Ichiishi, T. (1982b). Management versus ownership, I. *Internat. Econom. Rev.* **23**, 323–336.

Ichiishi, T. (1979). Management versus ownership, II. Unpublished mimeo.

Ichiishi, T., and Quinzii, M. (1981). Decentralization for the core of a production economy with increasing returns. Discussion Paper No. A230 0281, Laboratoire d' Econométrie de l' Ecole Polytechnique, Paris. Also available as Working Paper Series No. 81-3, College of Business Administration, Univ. of Iowa, Iowa City. To appear in *Internat. Econom. Rev.*

Ichiishi, T., and Schäffer, J. J. (1979). The $\mathscr{T}$-core of a game without sidepayments. Working Paper No. 9-79-80, Graduate School of Industrial Administration, Carnegie-Mellon University, Pittsburgh. (A revised version to appear in *Econom. Studies Quart.*)

Ichiishi, T., and Weber, S. (1978). Some theorems on the core of a non-sidepayment game with a measure space of players. *Internat. J. Game Theory* **7,** 95–112.

Kakutani, S. (1937). Ein Beweis des Satzes von M. Eidelheit über konvexe Menge. *Proc. Imperial Acad. Tokyo* **13,** 93–94.

Kakutani, S. (1941). A generalization of Brouwer's fixed-point theorem. *Duke Math. J.* **8,** 457–459.

Kalai, E., and Schmeidler, D. (1977). An admissible set occurring in various bargaining situations. *J. Econom. Theory* **14,** 402–411.

Kannai, Y. (1966). Values of games with a continuum of players. *Israel J. Math.* **4,** 54–58.

Kannai, Y. (1969). Countably additive measures in cores of games. *J. Math. Anal. Appl.* **27,** 227–240.

Kannai, Y. (1970). Continuity properties of the core of a market. *Econometrica* **38,** 791–815. (*See also* A correction. *Econometrica* **40** (1972), 955–958.)

Kannai, Y., and Mantel, R. (1978). Non-convexifiable Pareto sets. *Econometrica* **46,** 571–575.

Karamardian, S. (1971). Generalized complementarity problem. *J. Optim. Theory Appl.* **8,** 161–167.

Klee, V. L. (1951). Convex Sets in Linear Spaces. *Duke Math. J.* **18,** 443–466.

Knaster, B., Kuratowski, C., and Mazurkiewicz, S. (1929). Ein Beweis des Fixpunktsatzes für n-dimensionale Simplexe. *Fund. Math.* **14,** 132–137.

Kuratowski, C. (1933). *Topologie, I,* Warszawa: Universytet Warszawski. (English transl.: *Topology,* Vol. I. New York: Academic Press, 1966.)

Lange, O. (1942). The foundations of welfare economics. *Econometrica* **10,** 215–228.

Lemke, C. E., and Howson, J. T. (1964). Equilibrium points of bi-matrix games. *SIAM J. Appl. Math.* **12,** 413–423.

Lindenstrauss, J. (1966). A short proof of Liapounoff convexity-theorem. *J. Math. Mechan.* **15,** 971–972.

Lyapunov, A. (1940). Sur les fonctions-vecteurs complètement additives. *Bull. Acad. Sci. URSS, Sér. Math.* **4,** 465–478.

Mas-Colell, A. (1974). An equilibrium existence theorem without complete or transitive preferences. *J. Math. Econom.* **1,** 237–246.

Mas-Colell, A. (1977). Competitive and value allocations of large exchange economies. *J. Econom. Theory* **14,** 419–438.

McKenzie, L. W. (1959). On the existence of general equilibrium for a competitive market. *Econometrica* **27,** 54–71. (*See also:* Some corrections. *Econometrica* **29** (1961), 247–248.)

Michael, E. (1956). Continuous selections. I. *Ann. of Math.* **63,** 361–382.

Moulin, H. (1979). Two and three person games: A local study. *Internat. J. Game Theory* **8,** 81–107.

Myerson, R. B. (1977). Graphs and cooperation in games. *Math. Oper. Res.* **2,** 225–229.

Nash, J. F., Jr. (1950). Equilibrium points in n-person game. *Proc. Nat. Acad. Sci. U.S.A.* **36,** 48–49.

Nash, J. F., Jr. (1951). Non-cooperative games. *Ann. of Math.* **54,** 286–295.

Negishi, T. (1960). Welfare economics and existence of an equilibrium for a competitive economy. *Metroeconomica* **12,** 92–97.

Neyman, A. (1977). Values for non-transferable utility games with a continuum of players. Tech. Rep. No. 351. School of Operations Research and Industrial Engineering, Cornell Univ., Ithaca, New York.
Nikaidô, H. (1956). On the classical multilateral exchange problem. *Metroeconomica* **8**, 135–145.
Nikaidô, H. (1968). *Convex Structures and Economic Theory*. New York: Academic Press.
Nikaidô, H., and Isoda, K. (1955). Note on noncooperative convex games. *Pacific J. Math.* **5**, 807–815.
Owen, G. (1972). Values of games without side payments. *Internat. J. Game Theory* **1**, 95–109.
Owen, G. (1976). Values of market games without side payments. Unpublished mimeo.
Prakash, P., and Sertel, M. R. (1974). On the existence of noncooperative equilibria in social systems. Discussion Paper No. 92, The Center for Mathematical Studies in Economics and Management Science, Northwestern Univ., Evanston, Illinois.
Richter, H. (1963). Verallgemeinerung eines in der Statistik benötigten Satzes der Masstheorie. *Math. Ann.* **150**, 85–90. (Correction in the same volume, pp. 440–441.)
Rockafellar, R. T. (1970). *Convex Analysis*. Princeton: Princeton Univ. Press.
Rosenmüller, J. (1971). On core and value. *Oper. Res. Verfahren* **9**, 84–101.
Rosenmüller, J. (1975). Large games without side payments. *Oper. Res. Verfahren* **20**, 107–128.
Roth, A. E. (1977). The Shapley value as a von Neumann–Morgenstern utility. *Econometrica* **45**, 657–664.
Roth, A. E. (1980). Values for games without sidepayments: Some difficulties with current concepts. *Econometrica* **48**, 457–465.
Scarf, H. (1967a). The core of an N person game. *Econometrica* **35**, 50–69.
Scarf, H. (1967b). The approximation of fixed points of a continuous mapping. *SIAM J. Appl. Math.* **15**, 1328–1342.
Scarf, H. (1971). On the existence of a cooperative solution for a general class of N-person games. *J. Econ. Theory* **3**, 169–181.
Scarf, H. (1973). *The Computation of Economic Equilibria*. New Haven: Yale Univ. Press.
Schmeidler, D. (1967). On balanced games with infinitely many players. Research Program in Game Theory and Mathematical Economics, Research Memorandum No. 28, Department of Mathematics, The Hebrew Univ. of Jerusalem.
Schmeidler, D. (1972). Cores of exact games, I. *J. Math. Anal. Appl.* **40**, 214–225.
Schmeidler, D. (1973). Equilibrium points of nonatomic games. *J. Statist. Phys.* **7**, 295–300.
Schmeidler, D. (1980). Walrasian analysis via strategic outcome functions. *Econometrica* **48**, 1585–1593
Shafer, W. (1980). On the existence and interpretation of value allocation. *Econometrica* **48**, 467–476.
Shafer, W., and Sonnenschein, H. (1975). Equilibrium in abstract economies without ordered preferences. *J. Math. Econom.* **2**, 345–348.
Shapley, L. S. (1953). A value for n-person games. In H. W. Kuhn and A. W. Tucker (Eds.), *Contributions to the Theory of Games*, Vol. II, pp. 307–317. Princeton: Princeton Univ. Press.

Shapley, L. S. (1964). Values of large games—VII: A general exchange economy with money. Memorandum RM-4248-PR. Rand Corporation, Santa Monica, California.
Shapley, L. S. (1967). On balanced sets and cores. *Naval Res. Logist. Quart.* **14,** 453–460.
Shapley, L. S. (1969). Utility comparison and the theory of games. In *La Décision*, pp. 251–263. Paris: Edition du Centre National de la Recherche Scientifique.
Shapley, L. S. (1971). Cores of convex games. *Internat. J. Game Theory* **1,** 11–26.
Shapley, L. S. (1973). On balanced games without side payments. In T. C. Hu and S. M. Robinson (Eds.), *Mathematical Programming*, pp. 261–290. New York: Academic Press.
Shubik, M. (1959). Edgeworth market games. In A. W. Tucker and R. D. Luce (Eds.), *Contributions to the Theory of Games*, Vol. IV, pp. 267–278. Princeton: Princeton Univ. Press.
Starr, R. M. (1969). Quasi-equilibria in markets with non-convex preferences. *Econometrica* **37,** 25–38.
Stone, M. H. (1946). *Convexity*, Lecture Notes prepared by Harley Flanders, Univ. of Chicago, Chicago, Illinois.
Tijs, S. H. (1981). Bounds for the core and the τ-value. In O. Moeschlin and D. Pallaschke (Eds.), *Game Theory and Mathematical Economics*, pp. 123–132. Amsterdam: North-Holland.
Todd, M. (1978). Lecture Notes, School of Operations Research and Industrial Engineering, Cornell Univ., Ithaca, New York.
Tychonoff, A. (1935). Ein Fixpunktsatz. *Math. Ann.* **111,** 767–776.
Uzawa, H. (1962). Walras' existence theorem and Brouwer's fixed-point theorem. *Econom. Studies Quart.* **13,** 59–62.
Vind, K. (1965). A theorem on the core of an economy. *Rev. Econom. Studies* **32,** 47–48.
von Neumann, J. (1928). Zur Theorie der Gesellschaftsspiele. *Math. Ann.* **100,** 295–320. (English transl.: On the theory of games of strategy. In A. W. Tucker and R. D. Luce (Eds.), *Contributions to the Theory of Games*, Vol. IV, pp. 13–42. Princeton: Princeton Univ. Press, 1959.)
von Neumann, J. (1937). Über ein ökonomisches Gleichungssystem and eine Verallgemeinerung des Brouwerschen Fixpunktsatzes. *Ergebnisse eines Mathematischen Seminars* **8,** 73–83. (English transl.: A model of general economic equilibrium. *Rev. Econom. Studies* **13** (1945–1946), 1–9.)
von Neumann, J., and Morgenstern, O. (1947). *Theory of Games and Economic Behavior*, 2nd ed. Princeton: Princeton Univ. Press.
Walkup, D. W., and Wets, R. J.-B. (1969). Lifting projections of convex polyhedra. *Pacific J. Math.* **28,** 465–475.
Weber, R. J. (1978). Probabilistic values for games. Unpublished mimeo., Yale Univ., New Haven, Connecticut.
Weber, S. (1981). Some results on the weak core of a non-side-payment game with infinitely many players. *J. Math. Econom.* **8,** 101–111.
Yamazaki, A. (1978). An equilibrium existence theorem without convexity assumptions. *Econometrica* **46,** 541–555.

Author Index

A

Anderson, Robert M., 111
Arrow, Kenneth J., 71, 73, 111
Artstein, Zvi, 24
Aubin, Jean-Pierre, 112, 146
Aumann, Robert J., 73, 105, 108, 110, 111, 128, 145, 146, 147, 148

B

Banach, Stefan, 28
Berge, Claude, 31
Bewley, Truman F., 111
Billera, Louis J., 106, 112
Bixby, R. E., 112
Bondareva, O. N., 102
Border, Kim C., 49
Brouwer, Luitzen Egbertus Jan, 47
Browder, Felix E., 50

C

Carathéodory, C., 8
Cellina, Arrigo, 43
Champsaur, Paul, 148
Cheng, Hsueh-Cheng, 148
Cournot, Antoine Augustin, 70

D

Debreu, Gerard, 4, 70, 71, 72, 73, 109
Delbaen, Freddy, 104, 105
Dubey, Pradeep, 145

E

Edgeworth, Francis Ysidro, 5, 89, 109

F

Fan, Ky, 50, 51, 52, 106
Farkas, J., 78
Fenchel, Werner, 7
Folkman, J. H., 24

G

Gale, David, 72
Gillies, Donald B., 102
Glicksberg, Irving L., 50
Green, Edward J., 49

H

Hahn, Frank H., 111
Hahn, H., 28, 41

Harsanyi, John C., 146
Hart, Sergiu, 131, 148
Hausdorff, Felix, 40
Hildenbrand, Werner, 108, 110, 111
Hoeffding, Wassily, 144
Howe, Roger, 24
Howson, J. T., 44, 105

I

Ichiishi, Tatsuro, 105, 106, 107, 111, 146
Isoda, Kazuo, 70

K

Kakutani, Shizuo, 17, 48
Kalai, Ehud, 112
Kannai, Yakar, 104, 106, 110, 145, 146
Karamardian, Stephan, 50
Klee, Victor, L., 27
Knaster, Bronislaw, 47
Krein, Mark, 22
Kuratowski, Casimir, 41, 47

L

Lange, Oskar, 73
Lemke, Carlton E., 44, 105
Lindenstrauss, Joram, 25
Lyapunov, A., 25

M

McKenzie, Lionel W., 72
Mantel, Rolf R., 146
Mas-Colell, Andreu, 72, 132, 148
Mazurkiewicz, Stefan, 47
Mertens, Jean-François, 110
Michael, Ernest, 43
Milman, D., 22
Minkowski, Hermann, 28, 78
Morgenstern, Oskar, 17, 68, 102
Moulin, Hervé, 73
Myerson, Roger B., 145

N

Nash, John F., Jr., 69
Negishi, Takashi, 72

Neyman, Abraham, 146
Nikaidô, Hukukane, 7, 70, 72
Nikodym, Otton Martin, 104

O

Owen, Guillermo, 146, 148

P

Peleg, Bezalel, 105
Perles, M., 147
Prakash, Prem, 70, 74

Q

Quinzii, Martine, 111

R

Radon, Johann, 104
Richter, Hans, 25
Rockaffellar, R. Tyrell, 7, 23
Rosenmüller, Joachim, 104, 108, 146
Roth, Alvin E., 145, 146

S

Scarf, Herbert E., 105, 108, 109, 112
Schäffer, Juan J., 107
Schmeidler, David, 70, 73, 104, 112
Sertel, Murat R., 70, 74
Shafer, Wayne, J., 70, 146
Shapley, Lloyd S., 24, 102, 105, 108, 128, 145, 146, 147
Shubik, Martin, 109
Sonnenschein, Hugo, 70
Sperner, E., 44
Starr, Ross M., 24, 73
Stone, Marshall H., 17

T

Tijs, Stef H., 146
Todd, Michael, 105
Tucker, Albert W., 59
Tychonoff, Alexander, 50

U

Uzawa, Hirofumi, 73

V

Vietoris, L., 34
Vind, Karl, 110
von Neumann, John, 17, 49, 68, 102

W

Walkup, David W., 103
Weber, Robert J., 146
Weber, Shlomo, 106, 108
Wets, Roger J.-B., 103

Y

Yamazaki, Akira, 73

Subject Index

A

Abstract economy; see Game, Pseudo-game
Acceptable payoff vector, 111
Admissible set, 112
Affine function, 24
Affine hull, 11
Affine independence, 9
Affine subspace, 10
Algebraic interior, 13
Algebraic relative interior, 13
Attainable state, 72
Aumann–Shapley proposition, 128

B

Balanced game
    Balanced non-side-payment game, 83, 107
    Balanced side-payment game, 81, 103
    Balanced subfamily, 81, 107
    Balancing coefficient, 81
    Strongly balanced non-side-payment game, 115

C

Canonical form, 103
Closed correspondence; see Continuity of a correspondence
Coalition, 2
Coalition structure, 84
Commodity bundle, 3
Completely labeled subsimplex, 44
Composite correspondence, 33
Comprehensive non-side-payment game, 84
Concave function, 3
Consistency of linear inequalities, 78
Consumption set, 3
Continuity of a correspondence
    Closedness, 34
    H-lower semicontinuity, 40
    H-upper semicontinuity, 40
    H–K-upper semicontinuity, 41
    Upper demicontinuity, 52
    V-lower semicontinuity, 33
    V-upper semicontinuity, 33
Convex combination, 8
Convex function, 28

Convex hull, 8
Convex set, 8
Core
   α-Core, 111
   β-Core, 112
   Core allocation, 87
   Core correspondence, 102, 104
   Core of a generalized non-side-payment game, 85
   Core of a non-side-payment game, 83, 106
   Core of a side-payment game, 81, 103
   ϵ-Core, 108
   $\mathcal{T}$-Core, 107
   Weak core, 108
Correspondence, 32

D

Descriptive concept, 58

E

Economy, 63
Edgeworth box diagram, 5
Edgeworth proposition, 89
Epigraph, 28
Equal treatment property, 89, 109, 116, 129
Equilibrium
   Competitive equilibrium of a pure exchange economy, 5, 62, 73
   Equilibrium of a bimatrix game, 57
   Nash equilibrium of a game in normal form, 57
   Social coalitional equilibrium of a society; *see* Social coalitional equilibrium
   Social equilibrium of an abstract economy, 60, 71
   Strong equilibrium of a game in normal form; *see* strong equilibrium
Extreme point, 21

F

Face of a simplex, 44
Facial space, 24

G

Game, Pseudo-game
   Abstract economy, 60, 71
   Bimatrix game, 57
   Convex game, 103, 121
   Exact game, 103
   Game in characteristic function form with side payments, 80, 103
   Game in characteristic function form without side payments, 83, 106
   Game in normal form, 56
   Game in strategic form, 56
   Society, 95
   Zero-sum, two-person game, 67
Graph, 28, 32

H

Hausdorff metric, 40
Hyperplane, 12

I

Increasing returns with respect to coalition size, 121
Initial endowment vector, 4
Inverse of a correspondence
   Inverse, 32
   Lower inverse, 32
   Upper inverse, 32

J

Joint continuity, 70

L

Law of large numbers, 144
Lemma of
   Artstein, 24
   Browder–Karamardian, 50
   Minkowski–Farkas, 78
   Sperner, 44
Linear accessibility, 15

M

Marginal worth vector, 120
Market participant, 62, 73
Minimax principle, 68
Minkowski function, 28
Monotone side-payment game, 120
Multivalued function, 32

N

Nonsatiation point, 4
Normal vector, 12
Normative concept, 58
Null player, 118

O

Optimal strategy, 68
Outcome function, 56

P

Pareto optimum
    Pareto optimum of a game in normal form, 58, 100
    Pareto optimum of an economy, 63
Path-following technique, 44, 105
Player, 2
Preference relation
    Closed preference relation, 3
    Complete preference relation, 3
    Convex preference relation, 4
    Monotone preference relation, 4
    Nonsatiated preference relation, 4
    Strictly convex preference relation, 4
    Transitive preference relation, 3
    Weakly convex preference relation, 3
Prisoner's dilemma, 59
Proper support point, 20
Pure exchange economy, 4
Purely competitive sequence of economies, 110

Q

Quasi-concave function, 2

R

Regular economy, 75
Relative interior, 15
Replica economy, 89, 109
Retract, 53
Reversible point, 73

S

Saddle point, 69
Separation principle, 17, 50
Simplex, 10
Simplicial partition, 44
Social coalitional equilibrium, 95
Social equilibrium; see Equilibrium
Society; see Game, Pseudo-game
Solution to linear equalities, 78
Stationary point, 50
Strictly concave function, 3
Strong equilibrium, 100
Strong separation, 21
Subgradient, 29
Sublinearity, 28
Supergradient, 133
Supporting hyperplane, 20

T

Theorem
    α-Core nonemptiness, 112
    Brouwer's fixed-point, 47, 53, 74
    Carathéodory's, 8, 25
    Coincidence, 51
    Competitive equilibrium existence, 62, 72, 73, 74
    Continuous selection, 53
    Core allocation existence, 88
    Core nonemptiness, 81, 83, 86, 103, 106, 115
    Duality of LP, 113
    Fundamental theorems of welfare economics, 63
    Hahn–Banach, 28
    Kakutani's fixed-point, 48, 52, 54
    K–K–M, 47, 49
    K–K–M–S, 82
    Krein–Milman, 22

λ-Transfer value existence, 124
Limit theorem of cores, 90, 109, 110, 111
Limit theorem of value allocations, 130, 147, 148
Lyapunov's, 25
Maximum, 37
Nash equilibrium existence, 57
Separation, 20
Shapley–Folkman, 24
Shapley value unique existence, 118
Social coalitional equilibrium existence, 96, 99
Social equilibrium existence, 60, 70, 71
Strong equilibrium existence, 101
Strong separation, 21
Subdifferentiability, 29
Support, 19
Symmetric value allocation existence, 129
$\mathcal{T}$-Core nonemptiness, 107
Value allocation existence, 126, 147
Topological limes inferior, 40
Topological limes superior, 40

U

Utility function, 4

V

Value
  Asymptotic value, 145
  λ-Transfer value, 123
  Owen value, 146
  Shapley value, 120
  Symmetrical value allocation, 129
  τ-Value, 146
  Value, 145
  Value allocation, 126, 147
Value of a zero-sum two-person game, 68
Vertex
  Vertex of a simplex, 10
  Vertex of a simplicial partition, 44
Vietoris topology, 34

ECONOMIC THEORY, ECONOMETRICS, AND MATHEMATICAL ECONOMICS

Consulting Editor: Karl Shell

UNIVERSITY OF PENNSYLVANIA
PHILADELPHIA, PENNSYLVANIA

Franklin M. Fisher and Karl Shell. The Economic Theory of Price Indices: *Two Essays on the Effects of Taste, Quality, and Technological Change*

Luis Eugenio Di Marco (Ed.). International Economics and Development: *Essays in Honor of Raúl Presbisch*

Erwin Klein. Mathematical Methods in Theoretical Economics: *Topological and Vector Space Foundations of Equilibrium Analysis*

Paul Zarembka (Ed.). Frontiers in Econometrics

George Horwich and Paul A. Samuelson (Eds.). Trade, Stability, and Macroeconomics: *Essays in Honor of Lloyd A. Metzler*

W. T. Ziemba and R. G. Vickson (Eds.). Stochastic Optimization Models in Finance

Steven A. Y. Lin (Ed.). Theory and Measurement of Economic Externalities

David Cass and Karl Shell (Eds.). The Hamiltonian Approach to Dynamic Economics

R. Shone. Microeconomics: *A Modern Treatment*

C. W. J. Granger and Paul Newbold. Forecasting Economic Time Series

Michael Szenberg, John W. Lombardi, and Eric Y. Lee. Welfare Effects of Trade Restrictions: *A Case Study of the U.S. Footwear Industry*

Haim Levy and Marshall Sarnat (Eds.). Financial Decision Making under Uncertainty

Yasuo Murata. Mathematics for Stability and Optimization of Economic Systems

Alan S. Blinder and Philip Friedman (Eds.). Natural Resources, Uncertainty, and General Equilibrium Systems: *Essays in Memory of Rafael Lusky*

Jerry S. Kelly. Arrow Impossibility Theorems

Peter Diamond and Michael Rothschild (Eds.). Uncertainty in Economics: *Readings and Exercises*

Fritz Machlup. Methodology of Economics and Other Social Sciences

Robert H. Frank and Richard T. Freeman. Distributional Consequences of Direct Foreign Investment

Elhanan Helpman and Assaf Razin. A Theory of International Trade under Uncertainty

Edmund S. Phelps. Studies in Macroeconomic Theory, Volume 1: *Employment and Inflation.* Volume 2: *Redistribution and Growth.*

Marc Nerlove, David M. Grether, and José L. Carvalho. Analysis of Economic Time Series: *A Synthesis*

Thomas J. Sargent. Macroeconomic Theory

Jerry Green and José Alexander Scheinkman (Eds.). General Equilibrium, Growth and Trade: *Essays in Honor of Lionel McKenzie*

Michael J. Boskin (Ed.). Economics and Human Welfare: *Essays in Honor of Tibor Scitovsky*

Carlos Daganzo. Multinomial Probit: *The Theory and Its Application to Demand Forecasting*

L. R. Klein, M. Nerlove, and S. C. Tsiang (Eds.). Quantitative Economics and Development: *Essays in Memory of Ta-Chung Liu*

Giorgio P. Szegö. Portfolio Theory: *With Application to Bank Asset Management*

M June Flanders and Assaf Razin (Eds.). Development in an Inflationary World

Thomas G. Cowing and Rodney E. Stevenson (Eds.). Productivity Measurement in Regulated Industries

Robert J. Barro (Ed.). Money, Expectations, and Business Cycles: *Essays in Macroeconomics*

Ryuzo Sato. Theory of Technical Change and Economic Invariance: *Application of Lie Groups*

Iosif A. Krass and Shawkat M. Hammoudeh. The Theory of Positional Games: *With Applications in Economics*

Giorgio Szegö (Ed.). New Quantitative Techniques for Economic Analysis

John M. Letiche (Ed.). International Economic Policies and Their Theoretical Foundations: A Source Book

Murray C. Kemp (Ed.). Production Sets

Andreu Mas-Colell (Ed.). Noncooperative Approaches to the Theory of Perfect Competition

Jean-Pascal Benassy. The Economics of Market Disequilibrium

Tatsuro Ichiishi. Game Theory for Economic Analysis

In preparation

David P. Baron. The Export-Import Bank: *An Economic Analysis*